BIOLOGY

The Dynamics of Life

Reinforcement and Study Guide

Student Edition

New York, New York   Columbus, Ohio   Woodland Hills, California   Peoria, Illinois

A GLENCOE PROGRAM

BIOLOGY: THE DYNAMICS OF LIFE

Student Edition
Teacher Wraparound Edition
Laboratory Manual, SE and TE
Reinforcement and Study Guide, SE and TE
Content Mastery, SE and TE
Section Focus Transparencies and Masters
Reteaching Skills Transparencies and Masters
Basic Concepts Transparencies and Masters
BioLab and MiniLab Worksheets
Concept Mapping
Chapter Assessment
Critical Thinking/Problem Solving
Spanish Resources
Tech Prep Applications
Biology Projects
Computer Test Bank Software and Manual
  WINDOWS/MACINTOSH
Lesson Plans
Block Scheduling
Inside Story Poster Package
Science and Technology Videodisc Series, Teacher Guide
English/Spanish Audiocassettes
MindJogger Videoquizzes
Interactive CD-ROM
Videodisc Program
Glencoe Science Professional Series:
  Exploring Environmental Issues
  Performance Assessment in the Biology Classroom
  Alternate Assessment in the Science Classroom
  Cooperative Learning in the Science Classroom

Send all inquiries to:
Glencoe/McGraw-Hill
8787 Orion Place
Columbus, OH 43240

ISBN 0-02-828247-7
Printed in the United States of America.
3 4 5 6 7 8 9 10 047 08 07 06 05 04 03 02 01

Contents

To the Student

This *Reinforcement and Study Guide* for ***Biology: The Dynamics of Life*** will help you learn more easily from your textbook. Each textbook chapter has four study guide pages of questions and activities for you to complete as you read the text. The study guide pages are divided into sections that match those in your text. Each BioDigest in your textbook has two study guide pages to complete.

You will find that the directions in the *Reinforcement and Study Guide* are simply stated and easy to follow. Sometimes you will be asked to answer questions. Other times, you will be asked to label a diagram or complete a table. By completing the study guide, you will gain a better understanding of the concepts presented in the text. These sheets also will prove helpful when studying for a test.

Name _______________ Date _______ Class _______

The Study of Life

Reinforcement and Study Guide

Section 1.1 What Is Biology?

In your textbook, read about the science of biology.

Answer the following questions.

1. What is the primary focus of all biological studies?

____

____

2. What is meant by the statement, "Living things do not exist in isolation"?

____

____

In your textbook, read about why biologist study the diversity of life.

For each of the statements below, write <u>true</u> or <u>false</u>.

_______________ **3.** People study biology only if they are planning to become biologists.

_______________ **4.** By studying biology, you can better appreciate the great diversity of species on Earth and the way each species fits into the dynamic pattern of life on the planet.

_______________ **5.** The study of biology includes the investigation of interactions among species.

In your textbook, read about the characteristics of living things.

Complete each statement.

6. To be considered _______________, something must exhibit all of the _______________ of life.

7. _______________ is another word for "living thing."

8. Every living thing, from simple, single-celled organisms to complex, multicellular plants and animals, is made up of parts that function together in an orderly living _______________.

Read each of the following statements. If it describes the process of reproduction, write <u>yes</u>. If not, write <u>no</u>.

__________ **9.** New leaves appear on a tree in spring.

__________ **10.** An amoeba divides in half.

__________ **11.** A bean plant produces seeds in long pods.

__________ **12.** Pollen grains are released from a flower.

__________ **13.** A starfish produces a new arm after losing one to a predator.

Chapter 1

The Study of Life, *continued*

Reinforcement and Study Guide

Section 1.1 What Is Biology?, continued

Circle the letter of the choice that best completes the statement.

14. A species is defined as a group of similar-looking organisms that
a. undergo similar developmental changes. **b.** can interbreed.
c. can interbreed and produce fertile offspring. **d.** reproduce in the same way.

15. Every organism begins life as a(n)
a. embryo. **b.** single cell. **c.** nucleus. **d.** fertilized egg.

16. A corn plant producing ears of corn is an example of
a. growth. **b.** reproduction. **c.** development. **d.** all of these.

17. If members of a species fail to reproduce successfully, the species
a. will eventually become extinct. **b.** will not develop normally.
c. will evolve into a new species. **d.** will remain unchanged.

Complete the table below by checking the correct column for each example.

Example	Stimulus	Response
18. The recess bell ringing at an elementary school		
19. Your mouth watering at the sight of food on a plate		
20. A sudden drop in air temperature		
21. A flu virus entering your body		
22. Getting butterflies in your stomach before giving a speech		

Answer the following questions.

23. Explain the concept of homeostasis.

____

____

____

24. What is an adaptation?

____

____

____

25. What is evolution?

____

____

In your textbook, read about observing and hypothesizing.

Answer the following questions.

1. What is meant by *scientific methods*? ____

____

2. What is a hypothesis? ____

____

3. How is a hypothesis tested? ____

In your textbook, read about experimenting.

For each item in Column A, write the letter of the matching item in Column B.

	Column A	Column B
______	**4.** A procedure that tests a hypothesis by collecting information under controlled conditions	**a.** dependent variable
______	**5.** In an experiment, the group in which all conditions are kept the same	**b.** experimental group
______	**6.** In an experiment, the group in which all conditions are kept the same except for the one being tested	**c.** independent variable
______	**7.** The condition that is changed by the experimenter	**d.** experiment
______	**8.** The condition being observed or measured in an experiment	**e.** control group

Use each of the terms below just once to complete the passage.

experimental results	**experiment(s)**	**hypothesis**	**laws**
scientific journals	**theory**	**valid**	**verify**

When **(9)** ________________ are reported in **(10)** ________________, other scientists may try to **(11)** ________________ the results by repeating the **(12)** ________________. Usually when a(n) **(13)** ________________ is supported by data from several scientists, it is considered **(14)** ________________. Over time, a hypothesis that is supported by many observations and experiments becomes a **(15)** ________________. Some well-established facts of nature, such as gravity, are recognized as **(16)** ________________.

Chapter 1 **The Study of Life,** *continued*

Reinforcement and Study Guide

Section 1.3 The Nature of Biology

In your textbook, read about kinds of research.

Complete the chart by checking the correct column for each example.

Example	Quantitative Research	Descriptive Research
1. Numerical data		
2. Field study of hunting behavior		
3. Thermometer, balance scale, stopwatch		
4. Testable hypothesis		
5. Measurements from controlled laboratory experiments		
6. Purely observational data		
7. Binoculars, tape recorder, camera		

Complete each statement.

8. In order for scientific research to be universally understood, scientists report measurements in the ______________________________, a modern form of the metric system.

9. This system of measurement is abbreviated ____________________.

10. This system is a ____________________ system in which measurements are expressed in multiples of ____________________ or ____________________ of a basic unit.

In your textbook, read about science and society.

Determine if the following statement is true. If it is not, rewrite the italicized part to make it true.

11. Ideas about the value of knowledge gained through scientific research come from a society's *social, ethical*, and *moral* concerns. ______________________________

12. Pure science is scientific research carried out *primarily to solve a specific environmental problem.*

____

13. *Technology* is the practical application of scientific research to improve human life and the world in which we live. ______________________________

14. A technological solution to a human problem can benefit humans but may also *cause a different, possibly serious, problem.* ______________________________

15. *Scientists* have the final say about how the results of scientific discoveries are applied.

____

Name ______________________ Date ______________ Class ______________

Reinforcement and Study Guide

What Is Biology?

In your textbook, read about characteristics of life.

Complete the following statements.

Biology is the study of **(1)** ______________________ and the **(2)** ______________________ among them. Biologists use a variety of **(3)** ______________________ methods to study the details of life.

For each item in Column A, write the letter of the matching item in Column B.

	Column A	Column B
________	**4.** The basic unit is the cell.	**a.** development
________	**5.** Maintenance of a stable internal environment	**b.** growth
________	**6.** Reaction to a change in the environment	**c.** homeostasis
________	**7.** Cell enlargement and division	**d.** organization
________	**8.** Changes in an organism that take place over time	**e.** reproduction
________	**9.** Transmission of heredity information from one generation to the next	**f.** response to stimulus

Using what you know about characteristics of life, determine if each of the following describes a living or nonliving thing.

______________ **10.** rust on a bucket

______________ **11.** an apple on a tree

______________ **12.** bacteria

______________ **13.** lightning

______________ **14.** a dinosaur fossil

______________ **15.** a wasp

What Is Biology, *continued*

In your textbook, read about scientific methods.

Decide if each of the following statements is true. If it is not, rewrite the italicized part to make it true.

16. Scientific methods include observation, hypothesis, experiment, and *theory*. ______________________

17. A statement that can be tested and presents a possible solution to a question is a *law*.

____

18. In a controlled experiment, two groups are tested and all conditions except *two* are kept the same for both groups. ______________________

19. A condition that remains the same for both groups is called the *independent variable*.

____

20. A condition that is changed by the experimenter in one group and not the other is called the *dependent variable*. ______________________

21. A scientific experiment can be conducted *only in a laboratory*. ______________________

22. A theory is a *law* that has been confirmed by many experiments. ______________________

Read each of the following statements. If it is a testable hypothesis, write <u>yes</u>. If it is not a testable hypothesis, write <u>no</u>.

______________ **23.** If a person exercises, his or her pulse rate will increase.

______________ **24.** Cats make better pets than dogs.

______________ **25.** When fertilizer is added to soil, plants grow taller.

Identify each of the two italicized items as either an independent or a dependent variable.

26. The *number of red blood cells* in a mouse's blood at *different levels of iron* in its diet

____

27. The *amount of starch formed* in a plant leaf for *different times* of exposure to light

____

Name Date Class

Chapter 2

Principles of Ecology

Reinforcement and Study Guide

Section 2.1 Organisms and Their Environment

In your textbook, read about what ecology is and about aspects of ecological study.

Use each of the terms below just once to complete the passage.

ecology	**biotic factors**	**nonliving**	**environments**	**atmosphere**
humans	**organisms**	**soil**	**biosphere**	**abiotic factors**

Living organisms in our world are connected to other **(1)** ______________________ in a variety of ways. The branch of biology called **(2)** ______________________ is the scientific study of interactions among organisms and their **(3)** ______________________, including relationships between living and **(4)** ______________________ things.

All living things on Earth can be found in the **(5)** ______________________, the portion of Earth that supports life. It extends from high in the **(6)** ______________________ to the bottom of the oceans. Many different environments can be found in the biosphere. All living organisms found in an environment are called **(7)** ______________________. Nonliving parts of an environment are called **(8)** ______________________. For example, whales, trees, and **(9)** ______________________ are biotic factors. Ocean currents, temperature, and **(10)** ______________________ are abiotic factors.

In your textbook, read about levels of organization in ecology.

For each item in Column A, write the letter of the matching item in Column B.

	Column A	Column B
________	**11.** A group of organisms of one species that interbreed and live in the same place at the same time	**a.** community
________	**12.** A collection of interacting populations	**b.** competition
________	**13.** Interactions among the populations and abiotic factors in a community	**c.** forest
________	**14.** Occurs between organisms when resources are scarce	**d.** population
________	**15.** A terrestrial ecosystem	**e.** ecosystem

Chapter 2 **Principles of Ecology,** *continued*

Reinforcement and Study Guide

Section 2.1 Organisms and Their Environment, continued

In your textbook, read about organisms in ecosystems.

For each statement below, write true or false.

__________ **16.** A habitat is the role a species plays in a community.

__________ **17.** Habitats may change.

__________ **18.** A niche is the place where an organism lives its life.

__________ **19.** A habitat can include only one niche.

__________ **20.** A species' niche includes how the species meets its needs for food and shelter.

__________ **21.** The centipedes and worms that live under a certain log occupy the same habitat but have different niches.

__________ **22.** It is an advantage for two species to share the same niche.

__________ **23.** Competition between two species is reduced when the species have different niches.

Complete the table below by writing the kind of relationship described on the left.

Relationships Among Organisms	
Description of Relationship	**Kind of Relationship**
24. Organisms of different species live together in a close, permanent relationship.	
25. One species benefits and the other species is neither benefited nor harmed by the relationship.	
26. One species benefits from the relationship at the expense of the other species.	
27. Both species benefit from the relationship.	

Name Date Class

Section 2.2 Nutrition and Energy Flow

In your textbook, read about how organisms obtain energy and about matter and energy flow in ecosystems.

Answer the questions below. Use the diagram of a food web to answer questions 1–7.

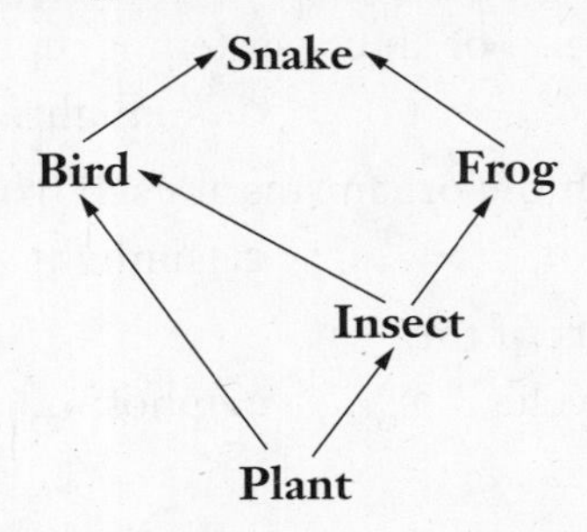

1. How many food chains make up the food web?

2. Which organism is an herbivore?

3. Which organism is an autotroph?

4. Which organism is a third-order heterotroph? To what trophic level does that organism belong?

5. Which organism is an omnivore?

6. Which organisms belong to more than one food chain?

7. Which organism belongs to more than one trophic level?

8. What are decomposers? From which trophic levels are the organisms that decomposers feed on?

9. What does a pyramid of energy show about the amount of energy available at different trophic levels of a food chain?

10. Why do different trophic levels have different amounts of energy?

Chapter **2** **Principles of Ecology,** ***continued***

Reinforcement and Study Guide

Section 2.2 Nutrition and Energy Flow, continued

In your textbook, read about cycles in nature.

Circle the letter of the choice that best completes the statement or answers the question.

11. Energy that is lost at each trophic level of an ecosystem is replenished by
a. heat. **b.** nutrients. **c.** sunlight. **d.** organisms.

12. Besides energy, what moves through the organisms at each trophic level of an ecosystem?
a. organisms **b.** nutrients **c.** sunlight **d.** cycles

13. Evaporation and condensation a part of the
a. carbon cycle. **b.** nitrogen cycle. **c.** phosphorus cycle. **d.** water cycle.

14. Plants lose water to the air through
a. condensation. **b.** photosynthesis. **c.** their roots. **d.** evaporation.

15. Animals lose water when they
a. breathe in. **b.** urinate. **c.** breathe out. **d.** both b and c.

16. The water in the atmosphere is returned to the earth by
a. precipitation. **b.** evaporation. **c.** photosynthesis. **d.** decomposition.

17. Autotrophs and heterotrophs use carbon molecules for energy and
a. photosynthesis. **b.** growth. **c.** decomposition. **d.** both a and b.

18. What do plants use in photosynthesis to make carbon molecules?
a. carbon dioxide **b.** carbohydrates **c.** fertilizer **d.** oxygen

19. Heterotrophs get carbon molecules by
a. making the molecules themselves. **b.** feeding on other organisms.
c. decaying. **d.** growing.

20. When decomposers break down the carbon molecules in dead organisms,
a. the dead organisms are converted to coal. **b.** oxygen is released.
c. carbon dioxide is released. **d.** carbon dioxide is converted to energy-rich carbon molecules.

21. Fertilizers provide plants with
a. nitrogen. **b.** carbon. **c.** water. **d.** oxygen.

22. Which of the following convert(s) nitrogen in the air into a form plants can use?
a. bacteria **b.** lightning **c.** sunlight **d.** both a and b

23. Plants use nitrogen to make
a. carbohydrates. **b.** nitrogen gas. **c.** proteins. **d.** both b and c.

24. An animal returns nitrogen to the environment when it
a. breathes. **b.** decomposes. **c.** urinates. **d.** both b and c.

25. Animals get phosphorus from
a. the air. **b.** eating plants. **c.** water. **d.** the soil.

26. Phosphorus in the soil comes from
a. rocks. **b.** decaying organisms. **c.** the air. **d.** both a and b.

Name ______________________ Date ________ Class ________

Chapter 3

Reinforcement and Study Guide

Communities and Biomes

Section 3.1 Communities

In your textbook, read about living in a community.

Determine if the statement is true. If it is not, rewrite the italicized part to make it true.

1. The *interactions* of abiotic and biotic factors result in conditions that are suitable for some organisms but not for others. ______________________

2. Food availability and temperature can be *biotic factors* for a particular organism. ______________________

3. A limiting factor is any biotic or abiotic factor that *promotes* the existence, numbers, reproduction, or distribution of organisms. ______________________

4. At high elevations where the soil is thin, vegetation is limited to *large, deep-rooted* trees.

____

5. Factors that limit one population in a community may also have *an indirect* effect on another population.

____

6. *Tolerance* is the ability of an organism to withstand fluctuations in biotic and abiotic environmental factors. ______________________

7. A population of deer would become *larger* as conditions move away from optimal toward either extreme of the deer's range of tolerance. ______________________

8. Different species may have different ranges of tolerance. ______________________

In your textbook, read about succession: changes over time.

Use each of the terms below just once to complete the passage.

climax	**primary**	**decades**	**succeed**
pioneer	**succession**	**species**	**slows down**

The natural changes and **(9)** ______________ replacements that take place in the communities of ecosystems are know as **(10)** ______________ . It can take **(11)** ______________ or even centuries for one community to **(12)** ______________ , or replace, another. When new sites of land are formed, as in a lava flow, the first organisms to colonize the new area are **(13)** ______________ species. This colonization is called **(14)** ______________ succession. The species inhabiting the area gradually change. Eventually, succession **(15)** ______________ and the community becomes more stable. Finally, a mature community that undergoes little or no change, called a **(16)** ______________ community, develops.

Section 3.1 Communities, continued

For each item in Column A, write the letter of the matching item in Column B.

	Column A	Column B
______	**17.** Sequence of community changes where soil is formed, allowing small, weedy plants to inhabit the area	**a.** a severe drought
______	**18.** Sequence of community changes occurring as a result of a natural disaster, such as a forest fire	**b.** primary succession
______	**19.** A stable, mature community with little or no succession occurring	**c.** amount of plant growth
______	**20.** An example of a biotic limiting factor affecting a community of organisms	**d.** secondary succession
______	**21.** An example of an abiotic limiting factor affecting a community of organisms	**e.** climax community

The statements below describe the secondary succession that occurred within an area of Yellowstone National Park. Number the events in the order in which they occurred.

______ **22.** Grasses, ferns, and pine seedlings inhabited the area.

______ **23.** Annual wildflowers grew from the bare soil.

______ **24.** A fire burned thousands of acres of land.

______ **25.** A climax community of lodgepole pines developed.

Name _______________ Date _______ Class _______

In your textbook, read about aquatic biomes: life in the water.

Complete each statement.

1. A large group of ecosystems sharing the same type of ______________________ is called a ______________________.

2. Biomes located in bodies of ______________________, such as oceans, lakes, and rivers, are called ______________________.

3. The water in marine biomes is ______________________.

4. Oceans contain the largest amount of ______________________, or living material, of any biome on Earth. Yet, most of the organisms are so ______________________ that they cannot be seen without magnification.

5. The ______________________ is that part of marine biomes shallow enough to be penetrated by sunlight.

6. Deep-water regions of marine biomes receiving no sunlight make up the ______________________.

Circle the letter of the response that best completes the statement.

7. If you followed the course of a river, it would eventually flow into
a. a lake. **b.** a stream. **c.** an ocean or a sea. **d.** a swamp.

8. The body of water where fresh water from a river mixes with salt water is called
a. an estuary. **b.** a shoreline. **c.** a sandbar. **d.** a sea.

9. Organisms living in intertidal zones have structures that protect them from
a. the dark. **b.** sunlight. **c.** wave action. **d.** temperature.

10. Life is abundant in photic zones because
a. there are no waves. **b.** the water is warm.
c. the water is clean. **d.** there are many nutrients.

11. The greatest number of organisms living in the photic zone of a marine biome are
a. dolphins. **b.** plankton. **c.** plants. **d.** sharks.

12. Few organisms live at the bottom of a deep lake because of the lack of
a. sunlight. **b.** space. **c.** plankton. **d.** bacteria.

In your textbook, read about terrestrial biomes.

Answer the following questions.

13. Which two abiotic factors generally determine the type of climax community that will develop in a particular part of the world?

____

14. In which terrestrial biome is the ground permanently frozen?

____

15. What are some adaptations that desert plants have developed?

____

____

____

16. Describe the three layers of a tropical rain forest, including organisms that live in each layer.

____

____

____

Write the name of each major terrestrial biome next to its description.

______________ **17.** Arid land with sparse, drought-resistant plants

______________ **18.** Largest terrestrial biome that supports small plants and grasses, but few trees

______________ **19.** Treeless land where only small plants and grasses grow during the long summer days

______________ **20.** Warm, wet land that supports many species of organisms

______________ **21.** Land with coniferous forests, peat swamps, and long, harsh winters

______________ **22.** Land populated with broad-leaved hardwood trees that lose their leaves annually

Name Date Class

Chapter 4

Population Biology

Reinforcement and Study Guide

Section 4.1 Population Dynamics

In your textbook, read about the principles of population growth.

Refer to Graphs A and B below. Answer the following questions.

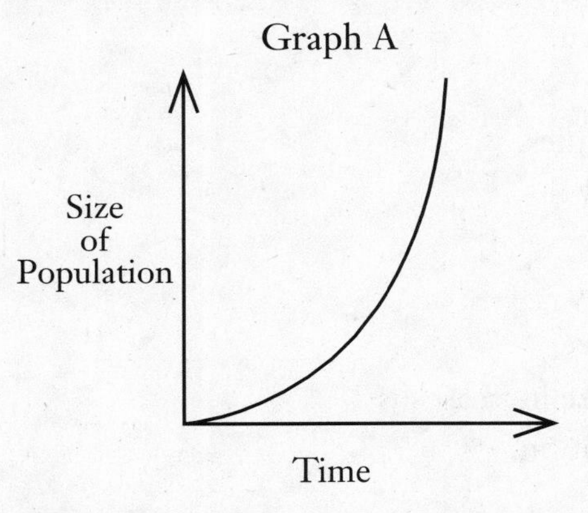

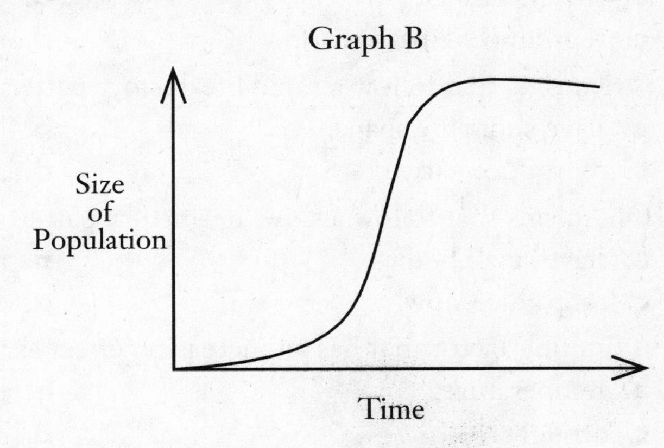

1. What type of population growth is shown in Graph A? Explain this type of growth.

____

____

2. Which graph shows the most likely growth of a squirrel population living in a forest? ______________

3. Which graph shows a population's growth under ideal conditions? ______________

4. Why don't populations of organisms grow indefinitely?

____

____

____

Use each of the terms below just once to complete the passage.

grows	**carrying capacity**	**below**	**births**
above	**under**	**deaths**	**exceed**

The number of organisms of one species that an environment can support is called its **(5)** ______________ . If the number of organisms in a population is **(6)** ______________ the environment's carrying capacity, births **(7)** ______________ deaths and the population **(8)** ______________ . If the number of organisms rises **(9)** ______________ the carrying capacity of the environment, **(10)** ______________ will exceed **(11)** ______________ . This pattern will continue until the population is once again at or **(12)** ______________ the carrying capacity.

Circle the letter of the choice that best completes the statement.

13. The most important factor that determines population growth is the organism's
- **a.** social pattern.
- **b.** carrying capacity.
- **c.** reproductive pattern.
- **d.** feeding pattern.

14. Organisms that follow a rapid life-history pattern
- **a.** have short life spans.
- **b.** have small bodies.
- **c.** reproduce early.
- **d.** all of the above

15. Organisms that follow a slow life-history pattern
- **a.** have small bodies.
- **b.** mature rapidly.
- **c.** reproduce slowly.
- **d.** all of the above

16. A limiting factor that has an increasing effect as population size increases is
- **a.** temperature.
- **b.** habitat disruption.
- **c.** drought.
- **d.** competition.

In your textbook, read about how organism interactions limit population size.

Answer the following.

17. The snowshoe hare is a primary source of food for the Canadian lynx. Explain how the lynx population size changes when the hare population increases.

____

____

18. Explain how the change in the lynx population size affects the hare population.

____

____

19. What is the relationship between the lynx and the hare called?

____

____

20. When does competition decrease the size of a population?

____

____

21. What can cause an organism to exhibit stress, and what symptoms of stress can lead to a decrease in population size?

____

____

Chapter 4 **Population Biology,** *continued*

Reinforcement and Study Guide

Section 4.2 Human Population Growth

In your textbook, read about demographic trends.

Determine if the statement is true. If it is not, rewrite the italicized part to make it true.

1. Looking at *past* population trends is a good way to predict the future of human populations.

2. Demography is the study of population *health* characteristics.

3. Worldwide human populations have *decreased* exponentially over the past few centuries.

4. Humans are able to *increase* environmental effects on the human population through controlling disease, eliminating competing organisms, and increasing food production.

5. To tell whether a population is *growing*, you must know the difference between the birthrate and the death rate.

6. The death rate is *decreasing* in the United States.

7. The birthrate is *increasing* in the United States.

8. *Birthrate* is the number of offspring a female produces during her reproductive years.

9. In the United States, families are now *smaller* than in previous decades.

10. Birthrates and death rates of countries around the world *are basically the same*.

11. If a country has a high death rate, it *may also* have a high birthrate.

12. If a country has a low death rate and a *high* birthrate, it will grow slowly, if at all.

Name ______________________ Date ______________ Class ______________

Chapter 4 **Population Biology,** *continued*

Reinforcement and Study Guide

Section 4.2 Human Population Growth, continued

For each statement in Column A, write the letter of the item in Column B that completes the statement correctly.

	Column A	Column B
______	**13.** Population growth will change if the largest ______________ of a population is in its post-reproductive years.	**a.** age structure
______	**14.** The proportions of a population that are at different ______________ make up its age structure.	**b.** stable
______	**15.** If you know a population has a large group of individuals in their pre-reproductive years, you would predict that the population's growth will be ______________ .	**c.** proportion
______	**16.** If the proportions of a population at different age levels are fairly equal, the population will be ______________ .	**d.** fertility
______	**17.** The population growth of a country depends on its birthrate, death rate, and ______________ rate.	**e.** rapid
______	**18.** To make predictions about the growth of a population, demographers must know its ______________ .	**f.** age levels

Complete each statement.

19. ______________________________ is the movement into and out of populations.

20. ______________________________ is the movement of humans into a population.

21. ______________________________ is the movement of humans from a population.

22. Immigration and emigration of people have no effect on total ______________________________ population.

23. Immigration and emigration of people affect ______________________________ population growth rates.

24. Suburban growth due to ______________________________ has placed stress on schools and various public services.

Name Date Class

Chapter 5

Biological Diversity and Conservation

Reinforcement and Study Guide

Section 5.1 Vanishing Species

In your textbook, read about biological diversity.

Use the terms below just once to complete the passage. You will not use all the terms.

environments	**variety**	**greater**	**space**	**species**
biological diversity	**equator**	**less**	**decrease**	**increase**

(1) ______________________ refers to the **(2)** ______________________ of life in an area. Another word for biological diversity is biodiversity. The simplest measure of biodiversity is the number of **(3)** ______________________ that live in a certain area. The more species there are, the **(4)** ______________________ is the biodiversity of the area. Biodiversity on land tends to **(5)** ______________________ as you move toward the **(6)** ______________________. Biodiversity is greater on large islands than on small islands because large islands have more **(7)** ______________________ and a greater variety of **(8)** ______________________.

In your textbook, read about the importance of biodiversity.

For each statement below, write true or false.

______________ **9.** Biodiversity provides our world with beauty.

______________ **10.** The loss of a species from an ecosystem usually has no effect because of the presence of other species in the ecosystem.

______________ **11.** Biodiversity decreases the stability of ecosystems because more species are competing with each other.

______________ **12.** Increasing the biodiversity of an ecosystem may result in more niches.

______________ **13.** Diseases are more likely to spread in an ecosystem with high biodiversity than in an ecosystem with low biodiversity.

______________ **14.** A decrease in Earth's biodiversity may affect people's diets.

______________ **15.** Preserving diverse plant species may lead to the discovery of new drugs in the future.

In your textbook, read about the loss of biodiversity.

For each item in Column A, write the letter of the matching item in Column B.

	Column A	Column B
______	**16.** The number of members of a species is so low that there is a possibility of extinction.	**a.** passenger pigeon
______	**17.** This animal is an example of an endangered species.	**b.** threatened species
______	**18.** The population of a species begins declining rapidly.	**c.** black rhinoceros
______	**19.** This animal is an example of an extinct species.	**d.** African elephant
______	**20.** All members of a species have died, so the species no longer exists.	**e.** extinct species
______	**21.** This animal is an example of a threatened species.	**f.** endangered species

In your textbook, read about threats to biodiversity.

Complete the table by checking the correct column for each statement.

Statement	Habitat Loss	Habitat Fragmentation	Habitat Degradation
22. Animals have no migratory route.			
23. A rain forest is burned.			
24. A highway divides a forest.			
25. Acid precipitation leaches nutrients from the soil.			
26. Detergents and other chemicals pollute bodies of water.			
27. Coral is mined for building materials.			
28. The reduction of the ozone layer causes more ultraviolet radiation to reach Earth's surface.			

Circle the letter of the choice that best completes the statement.

29. When species lose their habitats, they may
- **a.** lack food.
- **b.** lack shelter.
- **c.** be in danger of becoming extinct.
- **d.** all of the above.

30. Habitat fragmentation often leads to
- **a.** increased species diversity within an area.
- **b.** larger habitats for species.
- **c.** decreased species diversity within an area.
- **d.** an increased food supply for species.

31. Different conditions along the boundaries of an ecosystem are called
- **a.** habitat fragmentation.
- **b.** edge effect.
- **c.** habitat loss.
- **d.** canopy effect.

32. The greatest source of air pollution is
- **a.** volcanic eruptions.
- **b.** forest fires.
- **c.** burning fossil fuels.
- **d.** CFCs.

33. Acid precipitation
- **a.** may decrease biodiversity on land.
- **b.** has no effect on biodiversity.
- **c.** may increase biodiversity in water.
- **d.** both a and c.

34. The reduction of the ozone layer is caused by
- **a.** burning fossil fuels.
- **b.** acid precipitation.
- **c.** heavy metals.
- **d.** CFCs.

35. Algal blooms in lakes
- **a.** are caused by acid precipitation.
- **b.** decrease the amount of oxygen in the lake when they decay.
- **c.** clog the gills of fish.
- **d.** both a and b.

36. When exotic species are introduced into an area, their populations may grow exponentially because the species
- **a.** are large.
- **b.** are predators.
- **c.** lack competitors and predators.
- **d.** are small.

37. The African elephant population was greatly reduced between 1970 and 1990 due to
- **a.** habitat degradation.
- **b.** excessive hunting.
- **c.** habitat loss.
- **d.** pollution.

Chapter 5

Biological Diversity and Conservation, *continued*

Reinforcement and Study Guide

Section 5.2 Conservation of Biodiversity

In your textbook, read about strategies of conservation biology.

Answer the following questions.

1. What is conservation biology?

2. How does the U.S. Endangered Species Act protect biodiversity?

3. How do nature preserves help protect biodiversity?

4. Why is it usually better to preserve one large area of land instead of a few smaller areas of land?

5. Why are habitat corridors used to connect different protected areas?

6. What caused the steady decline of the black-footed ferret population in Wyoming?

7. What efforts were made to increase the size of the black-footed ferret population?

8. How are seed banks useful in protecting biodiversity?

9. What are some problems of keeping endangered animals in captivity before reintroducing them to their original habitats?

Name Date Class

BioDigest **2**

Ecology

Reinforcement and Study Guide

In your textbook, read about ecosystems.

For each statement below, write true or false.

_______________ **1.** Organisms interact with the nonliving parts of their environments.

_______________ **2.** Relationships between organisms are abiotic factors in ecosystems.

_______________ **3.** In the carbon cycle, animals produce nutrients from carbon dioxide in the atmosphere.

_______________ **4.** Commensalism is a relationship in which one species benefits while the other species is neither helped nor harmed.

_______________ **5.** The temperature and precipitation in a certain land area influence the type of biome that is found there.

In your textbook, read about food for life.

Use the diagram on the right to answer questions 6–10.

6. Describe a food chain using organisms in the pyramid.

7. Which organisms are carnivores?

8. How many trophic levels are included in the pyramid?

9. Which trophic level has the smallest biomass?

10. How does the biomass of the autotrophs compare with the biomass of the herbivores?

Name _______________ Date _______ Class _______

BioDigest 2

Ecology, *continued*

Reinforcement and Study Guide

In your textbook, read about population size.

Use the terms below to complete the passage. You will not use all the terms.

carrying capacity	**species**	**maximum**	**limit**	**competition**
linear growth	**minimum**	**exceeds**	**food**	**exponential growth**

A population is the number of organisms of one **(11)** _______________ that live in a certain area. Under ideal conditions in which there are no factors that **(12)** _______________ the size of a population, a population shows **(13)** _______________. However, in the environment, the sizes of populations are influenced by various limiting factors, such as the availability of **(14)** _______________, water, space, and other resources. As population size increases, **(15)** _______________ for the resources increases. The **(16)** _______________ size of a population that an environment can support is the environment's **(17)** _______________ for that population. When a population **(18)** _______________ the carrying capacity, individuals are unable to meet all their needs and die.

In your textbook, read about succession and biodiversity.

Number the steps of succession below in the order in which they occur.

_______ **19.** Shade from grasses and shrubs provides protection for tree saplings.

_______ **20.** Pioneer species and other small plants are unable to grow in the shade and die.

_______ **21.** Grasses and bushes appear.

_______ **22.** A plot of farmland is abandoned.

_______ **23.** Tree saplings grow and increase the amount of shade in the area.

_______ **24.** Pioneer species, such as dandelions, take root in the soil.

Answer the following questions.

25. What effect does succession have on the biodiversity of ecosystems?

____

26. What human actions decrease the biodiversity of ecosystems?

____

____

____

Chapter 6

The Chemistry of Life

Reinforcement and Study Guide

Section 6.1 Atoms and Their Interactions

In your textbook, read about elements, atoms, and isotopes.

Determine if the statement is true. If it is not, rewrite the italicized part to make it true.

1. An element is a substance that *can be* broken down into simpler substances. ____________

2. On Earth, *90* elements occur naturally. ____________

3. Only four elements—*carbon, hydrogen, oxygen, and nitrogen*—make up more than 96 percent of the mass of a human. ____________

4. Each element is abbreviated by a one- or two-letter *formula*. ____________

5. Trace elements, such as iron and magnesium, are present in living things in *very large* amounts. ____________

6. The properties of elements are determined by *the structures of their atoms*. ____________

Label the parts of the atom. Use these choices:

energy level **electron** **neutron** **proton** **nucleus**

e^- **11.** ____________

7. ____________ p^+ **10.** ____________

n^0

8. ____________ **9.** ____________

Answer the following questions.

12. What is the maximum number of electrons in each of the following energy levels: first, second, third?

____

____

13. Boron has two isotopes, boron-10 and boron-11. Boron-10 has five protons and five neutrons. How many protons and neutrons does boron-11 have? Explain.

____

____

____

____

____

____

Chapter 6 **The Chemistry of Life,** *continued*

Reinforcement and Study Guide

Section 6.1 Atoms and Their Interactions, continued

In your textbook, read about compounds and bonding, chemical reactions, and mixtures and solutions.

Write the type of substance described. Use these choices: compound, element.

______________ **14.** H_2O, a liquid that no longer resembles either hydrogen or oxygen gas

______________ **15.** A substance that can be broken down in a chemical reaction

______________ **16.** Carbon, the substance represented by the symbol C

Complete the table by checking the correct column for each description.

Statement	Ionic Bond(s)	Covalent Bond(s)
17. Found in the compound NaCl		
18. Increases the stability of atoms		
19. Results in the formation of a molecule		
20. Is formed when atoms share electrons		

Fill in the blanks with the correct number of molecules to balance the chemical equation. Then answer the questions.

$$C_6H_{12}O_6 + \_\_ O_2 \longrightarrow \_\_ CO_2 + \_\_ H_2O$$

21. Why must chemical equations always balance?

____

____

22. Which number indicates the number of atoms of each element in a molecule of a substance.

____

23. When is a mixture not a solution?

____

____

24. What is the difference between an acid and a base?

____

____

Chapter 6 The Chemistry of Life, *continued*

Section 6.2 Water and Diffusion

In your textbook, read about water and its importance.

For each statement below, write true or false.

__________ **1.** In a water molecule, electrons are shared equally between the hydrogen atoms and oxygen atom.

__________ **2.** The attraction of opposite charges between hydrogen and oxygen forms a weak oxygen bond.

__________ **3.** Because of its polarity, water can move from the roots of a plant up to its leaves.

__________ **4.** Water changes temperature easily.

__________ **5.** Unlike most substances, water expands when it freezes.

Circle the letter of the choice that best completes the statement.

6. All objects in motion have
a. potential energy. **b.** heat energy. **c.** kinetic energy. **d.** random energy.

7. The first scientist to observe evidence of the random motion of molecules was
a. Brown. **b.** Darwin. **c.** Mendel. **d.** Hooke.

8. The net movement of particles from an area of higher concentration to an area of lower concentration is called
a. dynamic equilibrium. **b.** nonrandom movement.
c. concentration gradient. **d.** diffusion.

9. Diffusion occurs because of
a. nonrandom movement of particles. **b.** random movement of particles.
c. a chemical reaction between particles. **d.** chemical energy.

10. When a few drops of colored corn syrup are added to a beaker of pure corn syrup, the color will
a. move from low concentration to high concentration.
b. form a polar bond.
c. start to diffuse.
d. remain on the bottom of the beaker.

11. Diffusion can be accelerated by
a. decreasing the pressure. **b.** increasing the temperature.
c. decreasing the movement of particles. **d.** increasing the dynamic equilibrium.

12. When materials pass into and out of a cell at equal rates, there is no net change in concentration inside the cell. The cell is in a state of
a. dynamic equilibrium. **b.** metabolism. **c.** imbalance. **d.** inertia.

13. The difference in concentration of a substance across space is called
a. dynamic equilibrium. **b.** concentration gradient.
c. diffusion. **d.** Brownian movement.

Name Date Class

Section 6.3 Life Substances

In your textbook, read about the role of carbon in organisms.

For each of the following statements about carbon, write <u>true</u> or <u>false</u>.

______ **1.** Carbon atoms can bond together in straight chains, branched chains, or rings.

______ **2.** Large molecules containing carbon atoms are called micromolecules.

______ **3.** Polymers are formed by hydrolysis.

______ **4.** Cells use carbohydrates for energy.

Write each item below under the correct heading.

sucrose	glucose	starch	$C_6H_{12}O_6$
cellulose	glycogen	fructose	$C_{12}H_{22}O_{11}$

Monosaccharide
5.
6.
7.

Dissaccharide
8.
9.

Polysaccharide
10.
11.
12.

Complete the table by checking the correct column for each description.

Description	Lipids	Proteins	Nucleic Acids
13. Made up of nucleotides			
14. Most consist of three fatty acids bonded to a glycerol molecule			
15. DNA and RNA			
16. Contain peptide bonds			
17. Produce proteins			
18. Commonly called fats and oils			
19. Made up of amino acids			
20. Used for long-term energy storage, insulation, and protective coatings			
21. Contain carbon, hydrogen, oxygen, and nitrogen			

Name Date Class

A View of the Cell

Reinforcement and Study Guide

Section 7.1 The Discovery of Cells

In your textbook, read about the history of the cell theory.

For each statement in Column A, write the letter of the matching item in Column B.

	Column A	Column B
______	**1.** The first scientist to describe living cells as seen through a simple microscope	**a.** Schleiden
______	**2.** Uses two or more glass lenses to magnify either living cells or prepared slides	**b.** compound light microscope
______	**3.** A scientist who observed that cork was composed of tiny, hollow boxes that he called cells	**c.** electron microscope
______	**4.** A scientist who concluded that all plants are composed of cells	**d.** Schwann
______	**5.** A scientist who concluded that all animals are composed of cells	**e.** Hooke
______	**6.** The microscope that allowed scientists to view molecules	**f.** Leeuwenhoek

In your textbook, read about the two basic cell types.

Complete the table by checking the correct column for each statement.

Statement	Prokaryotes	Eukaryotes
7. Organisms that have cells lacking internal membrane-bound structures		
8. Do not have a nucleus		
9. Are either single-celled or made up of many cells		
10. Generally are single-celled organisms		
11. Organisms that have cells containing organelles		

Section 7.2 The Plasma Membrane

In your textbook, read about maintaining a balance.

Use each of the terms below just once to complete the passage.

glucose	**plasma membrane**	**homeostasis**
organism	**balance**	**selective permeability**

Living cells maintain a **(1)** __________ by controlling materials that enter and leave. Without this ability, the cell cannot maintain **(2)** ______________ and will die. The cell must regulate internal concentrations of water, **(3)** _____________ , and other nutrients and must eliminate waste products. Homeostasis in a cell is maintained by the **(4)** ____________________ , which allows only certain particles to pass through and keeps other particles out. This property of a membrane is known as **(5)** ______________________ . It allows different cells to carry on different activities within the same **(6)** ___________ .

In your textbook, read about the structure of the plasma membrane.

For each statement below, write <u>true</u> or <u>false</u>.

__________________ **7.** The structure and properties of the cell wall allow it to be selective and maintain homeostasis.

__________________ **8.** The plasma membrane is a bilayer of lipid molecules with protein molecules embedded in it.

__________________ **9.** A phospholipid molecule has a nonpolar, water-insoluble head attached to a long polar, soluble tail.

__________________ **10.** The fluid mosaic model describes the plasma membrane as a structure that is liquid and very rigid.

__________________ **11.** Eukaryotic plasma membranes can contain cholesterol, which tends to make the membrane more stable.

__________________ **12.** Transport proteins span the cell membrane, creating the selectively permeable membrane that regulates which molecules enter and leave a cell.

__________________ **13.** Proteins at the inner surface of the plasma membrane attach the membrane to the cell's support structure, making the cell rigid.

Chapter 7

A View of the Cell, *continued*

Reinforcement and Study Guide

Section 7.3 Eukaryotic Cell Structure

In your textbook, read about cellular boundaries; nucleus and cell control; assembly, transport, and storage in the cell; and energy transformers.

Complete the table by writing the name of the cell part beside its structure/function. A cell part may be used more than once.

Structure/Function	Cell Part
1. A membrane-bound, fluid-filled sac	
2. Closely stacked, flattened membrane sacs	
3. The sites of protein synthesis	
4. A folded membrane that forms a network of interconnected compartments in the cytoplasm	
5. The clear fluid inside the cell	
6. Organelle that manages cell functions in eukaryotic cell	
7. Contains chlorophyll, a green pigment that traps energy from sunlight and gives plants their green color	
8. Digest excess or worn-out cell part, food particles, and invading viruses or bacteria	
9. Small bumps located on the endoplasmic reticulum	
10. Provides temporary storage of food, enzymes, and waste products	
11. Firm, protective structure that gives the cell its shape in plants, fungi, most bacteria, and some protists	
12. Produce a usable form of energy for the cell	
13. Modifies proteins chemically, then repackages them	
14. Contains inner membranes arranged in stacks of membranous sacs called grana	
15. Plant organelles that store starches or lipids or that contain pigments	

Name Date Class

In your textbook, read about structures for support and locomotion.

Determine if the statement is true. If it is not, rewrite the italicized part to make it true.

16. Cells have a support structure within the *cytoplasm* called the cytoskeleton.

____

17. The *exoskeleton* is composed of thin, fibrous elements that form a framework for the cell.

____

18. *Microtubules* of the cytoskeleton are thin, hollow cylinders made of protein.

____

19. Cilia and flagella are cell surface structures that are adapted for *respiration*.

____

20. *Flagella* are short, numerous, hairlike projections from the plasma membrane.

____

21. Flagella are longer and *more* numerous than cilia.

____

22. In *multicellular* organisms, cilia and flagella are the major means of locomotion.

____

23. In *prokaryotic* cells, both cilia and flagella are composed of microtubules.

____

Write titles for each of the generalized diagrams and then label the parts. Use these choices: plant cell, animal cell, plasma membrane, chloroplast, small vacuole, large vacuole, cell wall.

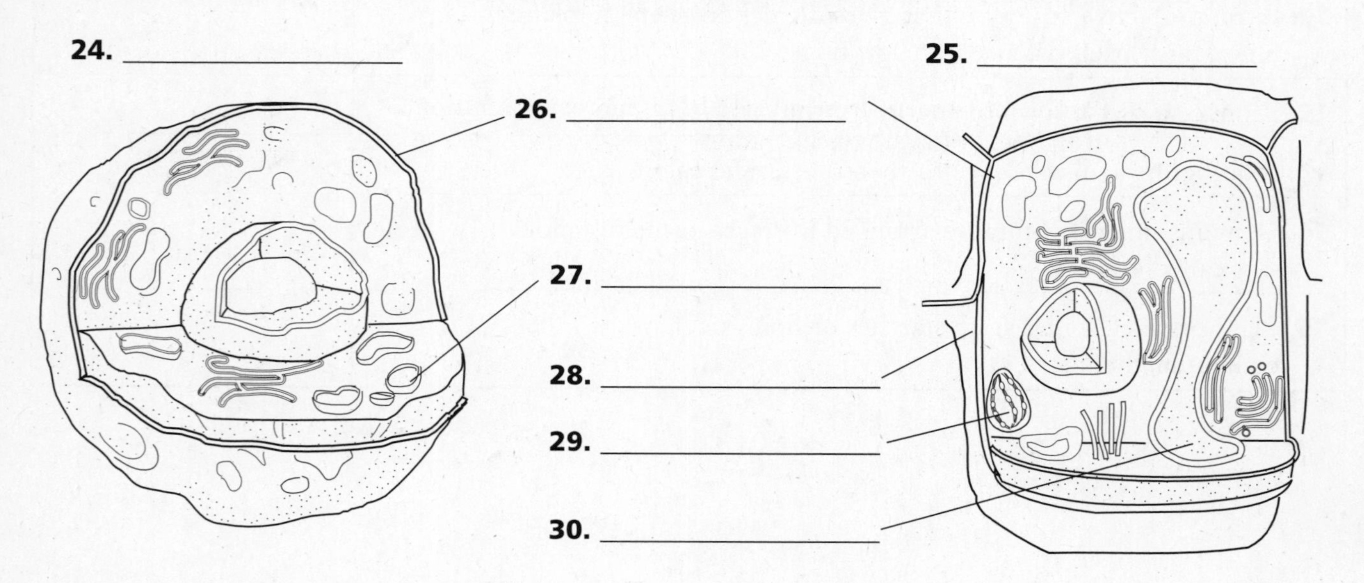

Name Date Class

Cellular Transport and the Cell Cycle

Reinforcement and Study Guide

Section 8.1 Cellular Transport

In your textbook, read about osmosis: diffusion of water.

Complete the table by checking the correct column for each statement.

Statement	Isotonic Solution	Hypotonic Solution	Hypertonic Solution
1. Causes a cell to swell			
2. Doesn't change the shape of a cell			
3. Causes osmosis			
4. Causes a cell to shrink			

In your textbook, read about passive transport and active transport.

For each item in Column A, write the letter of the matching item in Column B.

Column A

______ **5.** Transport protein that provides a tubelike opening in the plasma membrane through which particles can diffuse

______ **6.** Is used during active transport but not passive transport

______ **7.** Process by which a cell takes in material by forming a vacuole around it

______ **8.** Particle movement from an area of higher concentration to an area of lower concentration

______ **9.** Process by which a cell expels wastes from a vacuole

______ **10.** A form of passive transport that uses transport proteins

______ **11.** Particle movement from an area of lower concentration to an area of higher concentration

______ **12.** Transport protein that changes shape when a particle binds with it

Column B

a. energy

b. facilitated diffusion

c. endocytosis

d. passive transport

e. active transport

f. exocytosis

g. carrier protein

h. channel protein

Chapter 8

Cellular Transport and the Cell Cycle, *continued*

Reinforcement and Study Guide

Section 8.2 Cellular Growth and Reproduction

In your textbook, read about cell size limitations.

Determine if the statement is true. If it is not, rewrite the italicized part to make it true.

1. Most *living cells* are between 2 and 200 μm in diameter. ______________________
2. Diffusion of materials over long distance is *fast*. ______________________
3. If a cell doesn't have enough *DNA* to make all the proteins it needs, the cell cannot live. ______________________
4. As a cell's size increases, its volume increases much *slower* than its surface area. ______________________
5. If a cell's diameter doubled, the cell would require *two* times more nutrients and would have *two* times more wastes to excrete. ______________________

In your textbook, read about cell reproduction.

Use each of the terms below just once to complete the passage.

nucleus	**genetic material**	**chromosomes**	**packed**
identical	**chromatin**	**vanish**	**cell division**

The process by which two cells are produced from one cell is called **(6)** ______________. The two cells are **(7)** ______________ to the original cell. Early biologists observed that just before cell division, several short, stringy structures appeared in the **(8)** ______________. These structures seemed to **(9)** ______________ soon after cell division. These structures, which contain DNA and became darkly colored when stained, are now called **(10)** ______________. Scientists eventually learned that chromosomes carry **(11)** ______________, which is copied and passed on from generation to generation. Chromosomes normally exist as **(12)** ______________, long strands of DNA wrapped around proteins. However, before a cell divides, the chromatin becomes tightly **(13)** ______________.

In your textbook, read about the cell cycle and interphase.

Complete the table by checking the correct column for each statement.

Statement	Interphase	Mitosis
14. Cell growth occurs.		
15. Nuclear division occurs.		
16. Chromosomes are distributed equally to daughter cells.		
17. Protein production is high.		
18. Chromosomes are duplicated.		
19. DNA synthesis occurs.		
20. Cytoplasm divides immediately after this period.		
21. Mitochondria and other organelles are manufactured.		

In your textbook, read about the phases of mitosis.

Identify the following phases of mitosis. Use these choices: telophase, metaphase, anaphase, prophase. Then label the diagrams. Use these choices: sister chromatids, centromere, spindle fibers, centrioles.

22. ______________ **23.** ______________ **24.** ______________ **25.** ______________

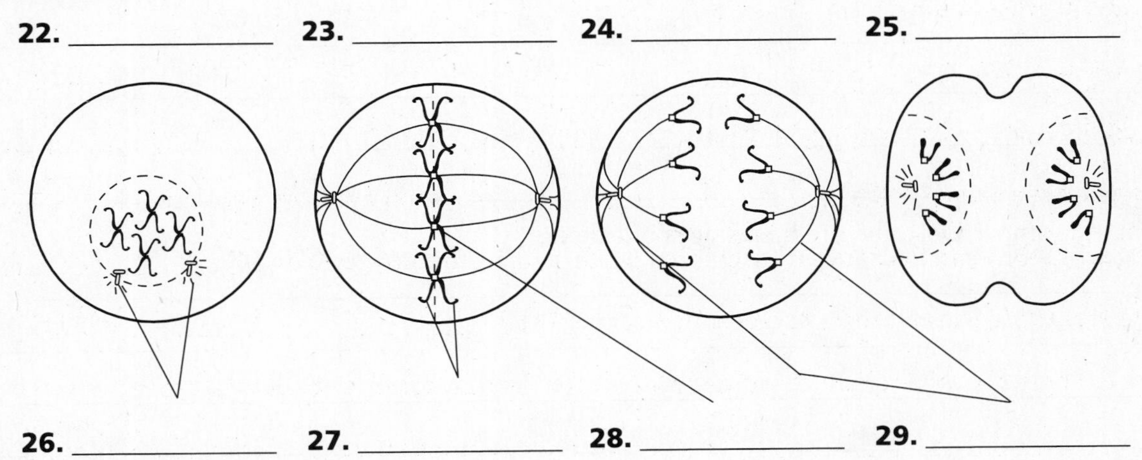

26. ______________ **27.** ______________ **28.** ______________ **29.** ______________

Answer the question.

30. How does mitosis result in tissues and organs?

____

____

____

Chapter 8

Cellular Transport and the Cell Cycle, *continued*

Reinforcement and Study Guide

Section 8.3 Control of the Cell Cycle

In your textbook, read about normal control of the cell cycle and cancer.

Answer the following questions.

1. In what ways do enzymes control the cell cycle?

2. What directs the production of these enzymes?

3. What can cause the cell cycle to become uncontrolled?

4. What can result when the cell cycle becomes uncontrolled?

5. What is the relationship between environmental factors and cancer?

6. What is a tumor? Describe the final stages of cancer.

7. Cancer is the second leading cause of death in the United States. What four types of cancer are the most prevalent?

Name Date Class

Chapter 9

Energy in a Cell

Reinforcement and Study Guide

Section 9.1 ATP in a Molecule

In your textbook, read about cell energy.

Use each of the terms below just once to complete the passage.

energy	**phosphate**	**adenosine**	**charged**
ATP	**chemical bonds**	**work**	**ribose**

To do biological **(1)** ______________________, cells require energy. A quick source of energy that cells use is the molecule **(2)** ______________________. The **(3)** ________________ in this molecule is stored in its **(4)** ______________________. ATP is composed of a(n) **(5)** ______________________ molecule bonded to a(n) **(6)** ______________________ sugar. Three **(7)** ______________________ molecules called **(8)** ______________________ groups are attached to the sugar.

In your textbook, read about forming and breaking down ATP and the uses of cell energy.

Examine the diagram below. Then answer the questions.

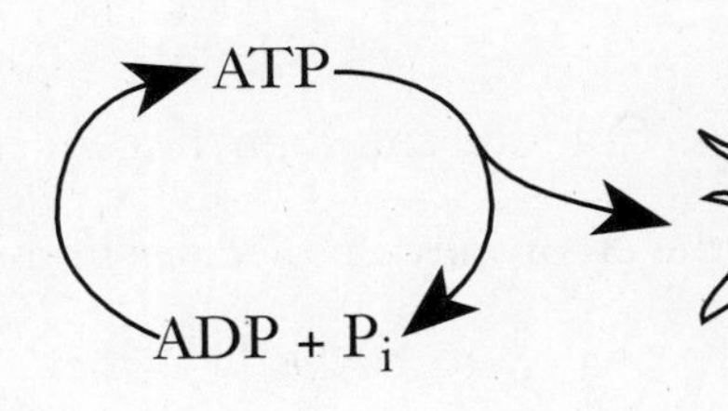

9. How is energy stored and released by ATP?

____

____

____

____

10. How do cells use the energy released from ATP?

____

____

____

Section 9.2 Photosynthesis: Trapping the Sun's Energy

In your textbook, read about trapping the sun's energy.

Determine if the statement is true. If it is not, rewrite the italicized part to make it true.

1. Photosynthesis is the process plants use to trap the sun's energy to make *glucose*.

____

2. ATP molecules are made during the *light-independent* reactions of photosynthesis.

____

3. *Carbon dioxide* gas is produced during photosynthesis.

____

4. The light-dependent reactions of photosynthesis take place in the membranes of the thylakoid discs in *mitochondria*.

____

5. The thylakoid membranes contain chlorophyll and other pigments that *absorb* sunlight.

____

In your textbook, read about the light-dependent reactions of photosynthesis.

Number the following steps of the light-dependent reactions in the order in which they occur.

_______ **6.** The energy lost by electrons as they pass through the electron transport chain is used to make ATP.

_______ **7.** The electrons pass from the chlorophyll to an electron transport chain.

_______ **8.** Sunlight strikes the chlorophyll molecules in the thylakoid membranes.

_______ **9.** $NADP^+$ molecules change to NADPH as they carry the electrons to the stroma of the chloroplast.

_______ **10.** The sunlight's energy is transferred to the chlorophyll's electrons.

_______ **11.** The electrons are passed down a second electron transport chain.

Answer the following questions.

12. How are the electrons that are lost by the chlorophyll molecules replaced?

____

____

13. How do plants produce oxygen during photosynthesis?

____

____

Name Date Class

Section 9.2 Photosynthesis: Trapping the Sun's Energy, continued

In your textbook, read about the light-independent reactions.

Circle the letter of the choice that best completes the statement or answers the question.

14. The Calvin cycle includes
- **a.** the light-dependent reactions.
- **b.** an electron transport chain.
- **c.** the light-independent reactions.
- **d.** photolysis.

15. The Calvin cycle takes place in the
- **a.** mitochondria.
- **b.** stroma.
- **c.** nucleus.
- **c.** thylakoid membrane.

16. What product of the light-dependent reactions is used in the Calvin cycle?
- **a.** oxygen
- **b.** carbon dioxide
- **c.** NADPH
- **d.** chlorophyll

17. What gas is used in the first step of the Calvin cycle?
- **a.** oxygen
- **b.** carbon dioxide
- **c.** hydrogen
- **d.** water

18. A carbon atom from carbon dioxide is used to change the five-carbon sugar RuBP into
- **a.** ATP.
- **b.** two molecules.
- **c.** PGA.
- **d.** a six-carbon sugar.

19. How many molecules of the three-carbon sugar PGA are formed?
- **a.** two
- **b.** one
- **c.** six
- **d.** three

20. ATP, NADPH, and hydrogen ions are used to convert PGA into
- **a.** PGAL.
- **b.** glucose.
- **c.** RuBP.
- **d.** carbon dioxide.

21. How many rounds of the Calvin cycle are needed to form one glucose molecule?
- **a.** one
- **b.** six
- **c.** two
- **d.** three

22. What two molecules leave the Calvin cycle and are combined to form glucose?
- **a.** RuBP
- **b.** PGA
- **c.** PGAL
- **d.** CO_2

23. Which molecule from the Calvin cycle is used to replenish the five-carbon sugar, RuBP, which is used at the beginning of the cycle?
- **a.** NADP
- **b.** CO_2
- **c.** PGA
- **d.** PGAL

Section 9.3 Getting Energy to Make ATP

In your textbook, read about the cellular respiration and fermentation.

Fill in the names of the molecules to complete the glycolysis reaction. Use these choices: 2PGAL, 4ATP, glucose, 2ADP, 2 pyruvic acid, 2NADH + $2H^+$. Then answer the questions.

Glycolysis

2ATP **2.** ____________

ENERGY

1. ____________ **3.** ____________

$4ADP + 4P_i$ **4.** ____________

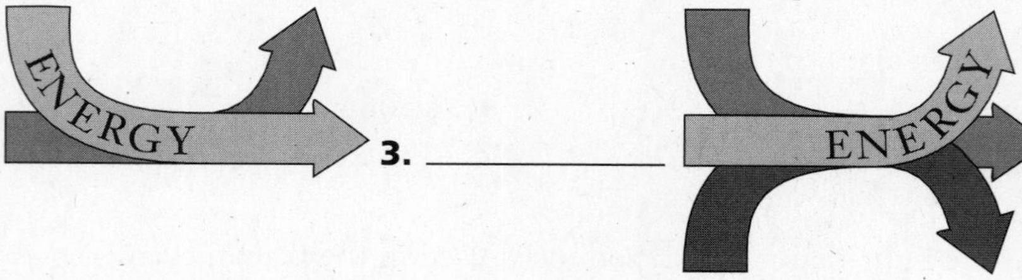

ENERGY

5. ____________

$2NAD^+$ **6.** ____________

7. What happens to pyruvic acid before entering the citric acid cycle?

____

8. What happens to the electrons carried by the NADH and $FADH_2$ molecules produced during the citric acid cycle?

____

____

9. During which stages of cellular respiration are ATP molecules formed?

____

10. Why is oxygen necessary for cellular respiration?

____

11. How is fermentation different from cellular respiration?

____

____

In your textbook, read about comparing photosynthesis and cellular respiration.

Answer the following question.

12. Describe two ways in which cellular respiration is the opposite of photosynthesis.

____

____

____

____

____

Name ______________________ Date __________ Class __________

BioDigest **3**

The Life of a Cell

Reinforcement and Study Guide

In your textbook, read about the chemistry of life.

Label the diagram below, using these choices:

atom **electron** **molecule** **neutron** **nucleus** **proton**

1. ______________________
2. ______________________
3. ______________________ 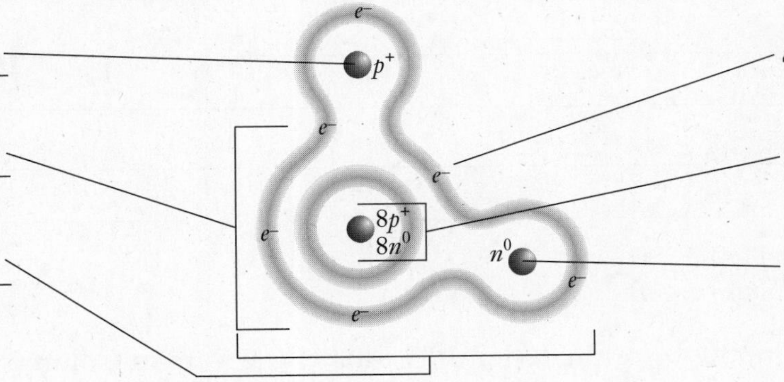
4. ______________________
5. ______________________
6. ______________________

In your textbook, read about eukaryotes, prokaryotes, and organelles.

Complete each statement.

7. Every cell is surrounded by a plasma ______________________.

8. ______________________ are organisms with cells that contain membrane-bound structures called organelles within the cell.

9. Organisms having cells without internal membrane-bound structures are called ______________________.

10. The plasma membrane is composed of a ______________________ with embedded proteins.

11. The ______________________ controls cell functions.

12. Ribosomes are organelles found in the cytoplasm that produce ______________________.

13. The ______________________ and Golgi apparatus transport and modify proteins.

14. Plant cells contain ______________________ that capture the sun's light energy so that it can be transformed into usable chemical energy.

15. A network of microfilaments and microtubules attached to the cell membrane give the cell ______________________.

16. ______________________ are long projections from the surface of the plasma membrane and move in a whiplike fashion to propel a cell.

BioDigest 3 **The Life of a Cell,** *continued*

Reinforcement and Study Guide

In your textbook, read about diffusion and osmosis.

Answer the following questions.

17. What is diffusion? ______________________________

18. What is osmosis? ______________________________

19. What is active transport? ______________________________

In your textbook, read about mitosis.

For each item in Column A, write the letter of the matching item in Column B.

	Column A	Column B
________	**20.** Duplicated chromosomes condense and mitotic spindles form on the two opposite ends of the cell.	**a.** anaphase
________	**21.** Chromosomes slowly separate to opposite ends of cells.	**b.** interphase
________	**22.** Chromosomes uncoil, spindle breaks down, and nuclear envelope forms around each set of chromosomes.	**c.** metaphase
________	**23.** Cells experience a period of intense metabolic activity prior to mitosis.	**d.** prophase
________	**24.** Chromosomes line up in center of cell.	**e.** telophase

In your textbook, read about energy in a cell.

Decide if each of the following statements is true. If it is not, rewrite the italicized part to make it true.

________ **25.** Adenosine triphosphate (ATP) is the most commonly used source of *protein* in a cell.

________ **26.** *Light-dependent* reactions convert energy into starch through the Calvin cycle.

________ **27.** *Mitochondria* convert food energy to ATP through a series of chemical reactions.

________ **28.** Glycolysis produces a net gain of two ATP for *every two molecules* of glucose.

Name Date Class

Chapter 10 Mendel and Meiosis

Section 10.1 Mendel's Laws of Heredity

In your textbook, read why Mendel succeeded.

Complete each statement.

1. Mendel was the first person to succeed in predicting how traits are ______________________ from generation to generation.

2. Mendel used ______________________ plants in his experiments.

3. In peas, both male and female sex cells—______________________—are in the same flower.

4. ______________________ occurs when the male gamete fuses with the female gamete.

5. Mendel used the process called ______________________ when he wanted to breed one plant with another.

6. Mendel carefully ______________________ his experiments and the peas he used.

7. Mendel studied only one ______________________ at a time and analyzed his data mathematically.

In your textbook, read about Mendel's monohybrid crosses.

Refer to the table of pea-plant traits on the right. Then complete the table on the left by filling in the missing information for each cross. The first one is done for you.

Parent Plants	F_1 generation	
	Offspring	Appearance
8. round × wrinkled *RR* × *rr*	*Rr*	round
9. yellow × green *YY* × *yy*		
10. axial × terminal *AA* × ______	*Aa*	
11. tall × short ______ × ______	*Tt*	
12. inflated × constricted ______ × *ii*		

Pea-Plant Traits		
Trait	Dominant	Recessive
seed shape	round (*R*)	wrinkled (*r*)
seed color	yellow (*Y*)	green (*y*)
flower position	axial (*A*)	terminal (*a*)
plant height	tall (*T*)	short (*t*)
pod shape	inflated (*I*)	constricted (*i*)

Chapter 10 **Mendel and Meiosis,** *continued*

Reinforcement and Study Guide

Section 10.1 Mendel's Laws of Heredity, continued

In your textbook, read about phenotypes and genotypes and Mendel's dihybrid crosses.

Determine if the statement is true. If it is not, rewrite the underlined part to make it true.

13. A pea plant with the genotype *TT* has the same phenotype as a pea plant with genotype <u>*tt*</u>. ________

14. When Mendel crossed true-breeding pea plants that had round yellow seeds with true-breeding pea plants that had wrinkled green seeds, <u>some</u> of the offspring had round yellow seeds because round and yellow were the dominant forms of the traits. ________

15. When Mendel allowed heterozygous F_1 plants that had round yellow seed to self-pollinate, he found that <u>some</u> of the F_2 plants had wrinkled green seeds. ________

16. The law of independent assortment states that <u>genes</u> for different traits are inherited independently of each other. ________

In your textbook, read about Punnett squares and probability.

The Punnett square below is for a dihybrid cross between pea plants that are heterozygous for seed shape (*Rr*) and seed color (*Yy*). Complete the Punnett square by recording the expected genotypes of the offspring. Then answer the questions.

	RY	*Ry*	*rY*	*ry*
RY				
Ry				
rY				
ry				

17. Use the chart on the previous page to determine the phenotypes of the offspring. Record the phenotypes below the genotypes in the Punnett square. Is an offspring produced by the cross more likely to have wrinkled seeds or round seeds? ________

18. What is the probability that an offspring will have wrinkled yellow seeds? ________

Name Date Class

Chapter 10

Mendel and Meiosis, *continued*

Reinforcement and Study Guide

Section 10.2 Meiosis

In your textbook, read about genes, chromosomes, and numbers.

Examine the table. Then answer the questions.

Chromosome Numbers of Some Common Organisms

Organism	Body Cell (2*n*)	Gamete (*n*)
Human	46	23
Garden pea	14	7
Fruit fly	8	4
Tomato	24	12
Dog	78	39
Chimpanzee	48	24
Leopard frog	26	13
Corn	20	10

1. What is the diploid number of chromosomes in corn?

____

2. What is the haploid number of chromosomes in corn?

____

3. Is the chromosome number related to the complexity of the organism?

____

4. How many pairs of chromosomes do humans have?

____

5. What process maintains a constant number of chromosomes within a species?

____

In your textbook, read about the phases of meiosis.

Label the diagrams below. Use these choices: Metaphase I, Metaphase II, Interphase, Telophase I, Telophase II, Anaphase I, Anaphase II, Prophase I, Prophase II.

6. ______________ **7.** ______________ **8.** ______________ **9.** ______________ **10.** ______________

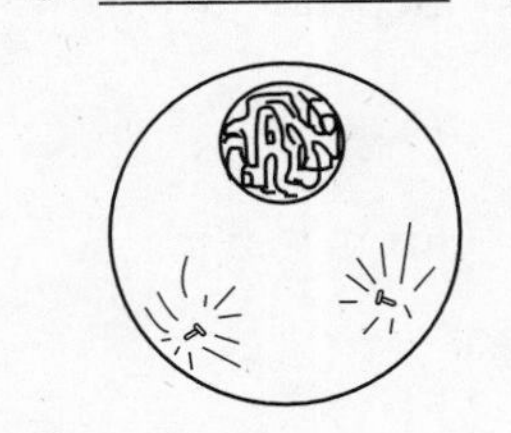

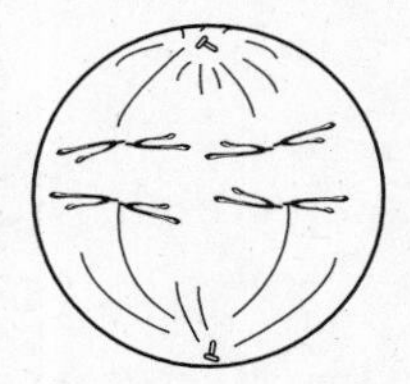
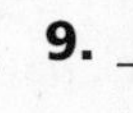
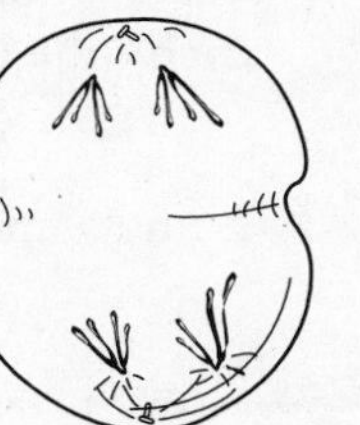
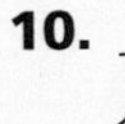

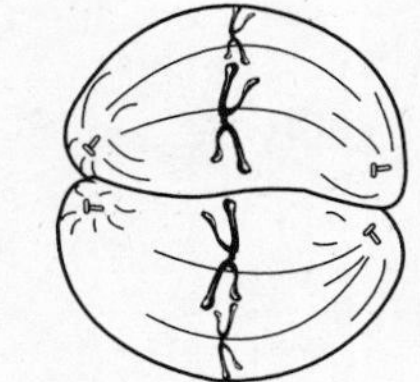
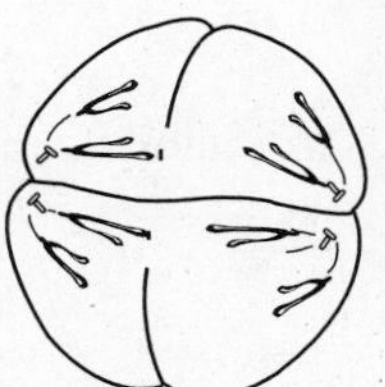
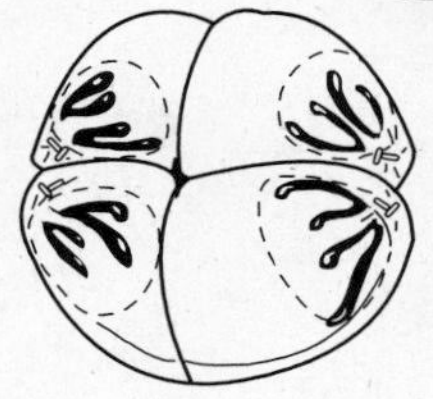

11. ______________ **12.** ______________ **13.** ______________ **14.** ______________

Section 10.2 Meiosis, continued

The following statements describe interphase and and meiosis I. Identify each phase. Then place them in sequential order using the numbers 1 through 5. Use 1 for the phase that occurs first and 5 for the phase that occurs last.

Statement	Name of Phase	Sequence
15. Homologous chromosomes line up at the equator in pairs.		
16. The cell replicates its chromosomes.		
17. Homologous chromosomes separate and move to opposite ends of the cell.		
18. The spindle forms, and chromosomes coil up and come together in a tetrad; crossing over may occur.		
19. Events occur in the reverse order from the events of prophase I. Each cell has only half the genetic information; however, another cell division is needed because each chromosome is still doubled.		

In your textbook, read about how meiosis provides for genetic variation and about mistakes in meiosis.

For each statement below, write true or false.

__________ **20.** Reassortment of chromosomes can occur during meiosis by crossing over or by independent segregation of homologous chromosomes.

__________ **21.** Genetic recombination is a major source of variation among organisms.

__________ **22.** The random segregation of chromosomes during meiosis explains Mendel's observation that genes for different traits are inherited independently of each other.

__________ **23.** Nondisjunction always results in a zygote with an extra chromosome.

__________ **24.** Down syndrome is a result of polyploidy.

__________ **25.** Mistakes in meiosis can occasionally be beneficial.

Name ____________________ Date ____________ Class ____________

Chapter 11

DNA and Genes

Reinforcement and Study Guide

Section 11.1 DNA: The Molecule of Heredity

In your textbook, read about what DNA is and the replication of DNA.

Label the diagram. Use these choices: nucleotide, deoxyribose, phosphate group, nitrogen base, hydrogen bonds, base pair.

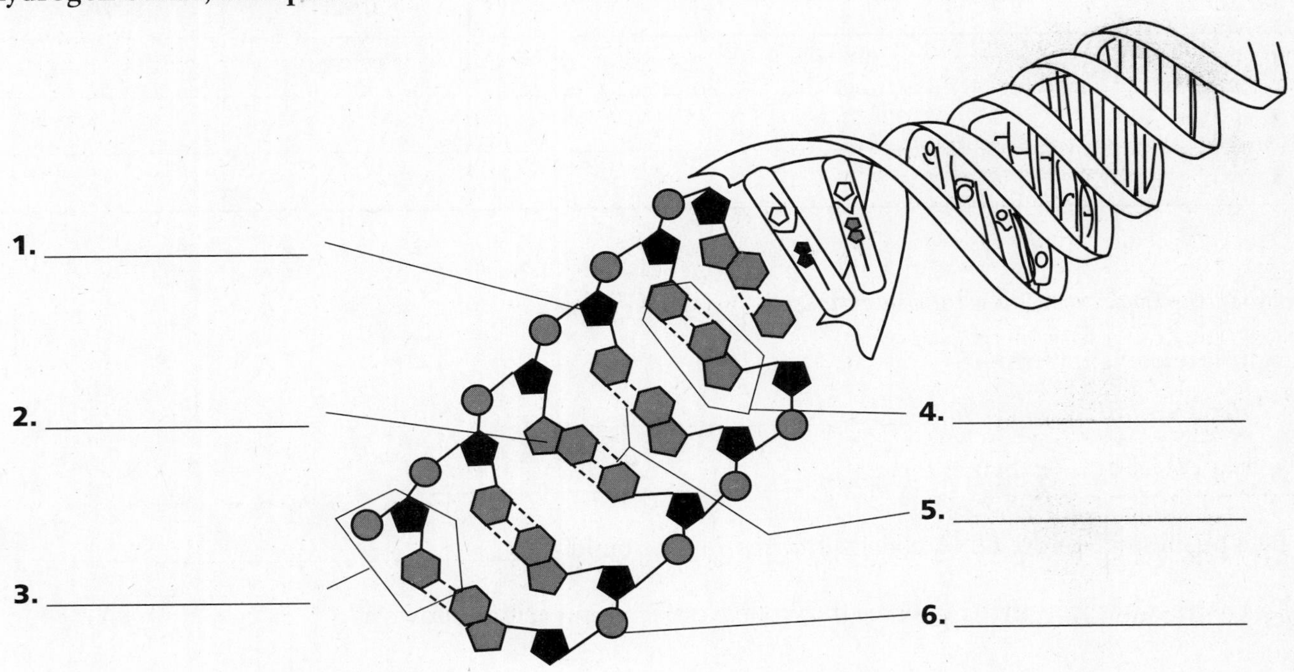

Complete each statement.

7. ____________________, guanine (G), cytosine (C), and thymine (T) are the four ____________________ in DNA.

8. In DNA, ____________________ always forms hydrogen bonds with guanine (G).

9. The sequence of ____________________ carries the genetic information of an organism.

10. The process of ____________________ produces a new copy of an organism's genetic information, which is passed on to a new cell.

11. The double-coiled shape of DNA is called a ____________________.

Chapter 11 **DNA and Genes,** *continued*

Reinforcement and Study Guide

Section 11.2 From DNA to Protein

In your textbook, read about genes and proteins and RNA.

Complete the chart on the three chemical differences between DNA and RNA.

Structure	DNA	RNA
1. strand of nucleotides		
2. sugar		
3. nitrogen base		

In your textbook, read about the genetic code.

Complete each statement.

4. Proteins are made up of ______________________ .

5. There are twenty different types of ______________________ .

6. The message of the DNA code is information for building ______________________ .

7. Each set of three nitrogen bases that codes for an amino acid is known as a ______________________ .

8. The amino acid ______________________ is represented by the mRNA codon ACA.

9. ______________________ and ______________________ are mRNA codons for phenylalanine.

10. There can be more than one ______________________ for the same amino acid.

11. For any one codon, there can be only one ______________________ .

12. The genetic code is said to be universal because a codon represents the same ______________________ in almost all organisms.

13. ______________________ , ______________________ , and ______________________ are stop codons.

14. ______________________ and ______________________ are amino acids that are each represented by only one codon.

Name Date Class

DNA and Genes, *continued*

Reinforcement and Study Guide

Section 11.2 From DNA to Protein, continued

In your textbook, read about transcription from DNA to mRNA.

Complete each statement.

15. Proteins are made in the cytoplasm of a cell, whereas DNA is found only in the ______________________________.

16. The process of making RNA from DNA is called ______________________________.

17. The process of transcription is similar to the process of DNA ______________________________.

18. ______________________________ carries information from the DNA in the nucleus out into the cytoplasm of the cell.

19. mRNA carries the information for making proteins to the ______________________________.

In your textbook, read about translation from mRNA to protein.

Label the diagram. Use these choices: transfer RNA (tRNA), amino acid, amino acid chain, codon, anticodon, messenger RNA (mRNA), ribosome.

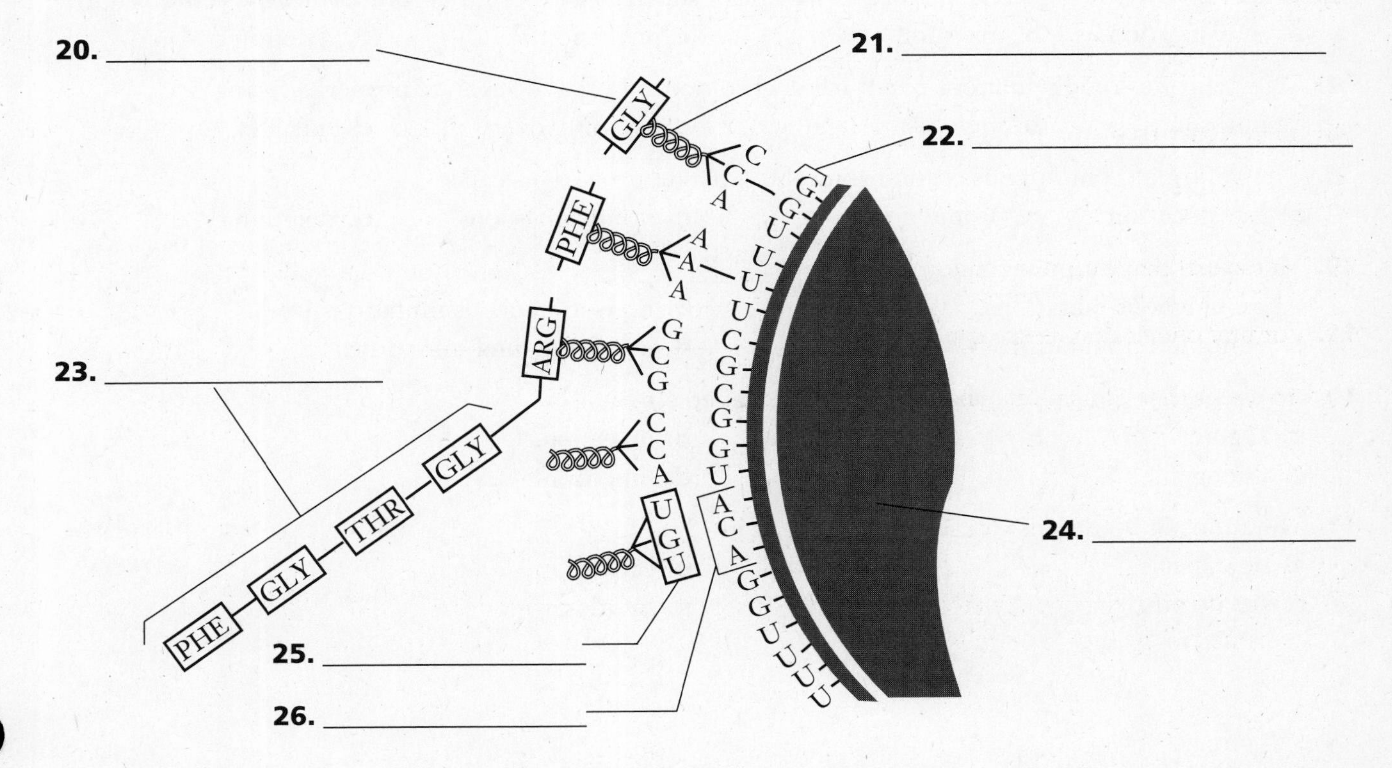

Chapter 11 **DNA and Genes,** *continued*

Reinforcement and Study Guide

Section 11.3 Genetic Changes

In your textbook, read about mutation: a change in DNA.

Circle the letter of the choice that best completes the statement.

1. A mutation is any mistake or change in the
a. cell. **b.** DNA sequence. **c.** ribosomes. **d.** nucleus.

2. A point mutation is a change in
a. several bases in mRNA. **b.** several bases in tRNA.
c. a single base pair in DNA. **d.** several base pairs in DNA.

3. A mutation in which a single base is added or deleted from DNA is called
a. a frame shift mutation. **b.** a point mutation. **c.** translocation. **d.** nondisjunction.

4. Chromosomal mutations are especially common in
a. humans. **b.** animals. **c.** bacteria. **d.** plants.

5. Few chromosome mutations are passed on to the next generation because
a. the zygote usually dies.
b. the mature organism is sterile.
c. the mature organism is often incapable of producing offspring.
d. all of the above.

6. When part of one chromosome breaks off and is added to a different chromosome, the result is a(n)
a. translocation. **b.** insertion. **c.** inversion. **d.** deletion.

7. Many chromosome mutations result when chromosomes fail to separate properly during
a. mitosis. **b.** meiosis. **c.** crossing over. **d.** linkage.

8. The failure of homologous chromosomes to separate properly is called
a. translocation. **b.** disjunction. **c.** nondisjunction. **d.** deletion.

9. Mutations that occur at random are called
a. spontaneous mutations. **b.** nonspontaneous mutations.
c. nonrandom mutations. **d.** environmental mutations.

10. An agent that can cause a change in DNA is called a(n)
a. zygote. **b.** inversion.
c. mutagen. **d.** mutation.

11. Mutations in body cells can sometimes result in
a. new species. **b.** cancer.
c. sterile offspring. **d.** hybrids.

Name Date Class

Chapter 12

Patterns of Heredity and Human Genetics

Reinforcement and Study Guide

Section 12.1 Mendelian Inheritance of Human Traits

In your textbook, read about making a pedigree.

Examine the pedigree to the right. Then answer the following questions.

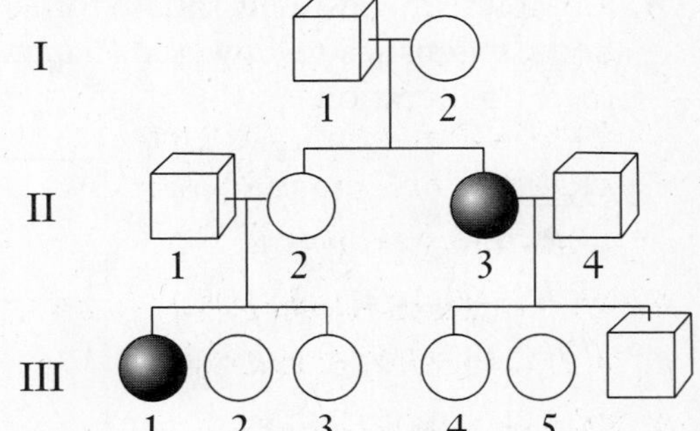

1. Is the trait being studied in the pedigree recessive or dominant? How do you know?

____

____

____

____

____

2. Are II-1 and II-2 carriers of the trait? How do you know?

____

____

3. What is the probability that II-1 and II-2 will produce an individual with the trait being studied? Draw a Punnett square to show your work.

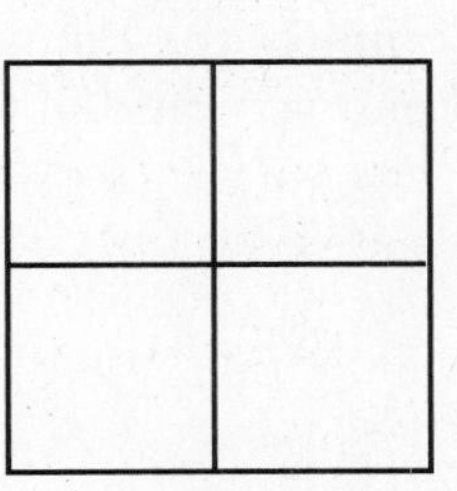

____

____

____

____

4. What is the likely genotype of II-4 for the trait being studied in the pedigree?

____

In your textbook, read about simple recessive heredity and simple dominant heredity.

For each item in Column A, write the letter of the matching item from Column B.

	Column A	Column B
______	**5.** Recessive disorder that results from the absence of an enzyme required to break lipids down	**a.** cystic fibrosis
______	**6.** Lethal genetic disorder caused by a dominant allele	**b.** simple dominant traits
______	**7.** Most common genetic disorder among white Americans	**c.** Tay-Sachs disease
______	**8.** Recessive disorder that results from the absence of an enzyme that converts one amino acid into another one	**d.** Huntington's disease
______	**9.** Tongue curling and Hapsburg lip	**e.** phenylketonuria

Name Date Class

Chapter 12

Patterns of Heredity and Human Genetics, *continued*

Reinforcement and Study Guide

Section 12.2 When Heredity Follows Different Rules

In your textbook, read about complex patterns of inheritance.

Answer the following questions.

1. Complete the Punnett square for a cross between a homozygous red-flowered snapdragon (*RR*) and a homozygous white-flowered snapdragon (*R'R'*). Give the genotype and phenotype of the offspring in the F_1 generation.

Key
RR - red
R'R' - white
RR' - pink

F_1

genotype: ____________________

phenotype: ____________________

2. When traits are inherited in an incomplete dominance pattern, what is true of the phenotype of the heterozygotes?

____

3. Complete the Punnett square for a cross between two pink-flowered (*RR'*) F_1 plants. Give the phenotype ratio of the offspring in the F_2 generation.

F_2

phenotype ratio: ____________________

4. In what type of inheritance are both alleles expressed equally?

____

5. Complete the Punnett square for a cross between a black chicken (*BB*) and a white chicken (*WW*). Give the phenotype of the offspring in the F_1 generation.

Key
BB - black
WW - white
BW - checkered

F_1

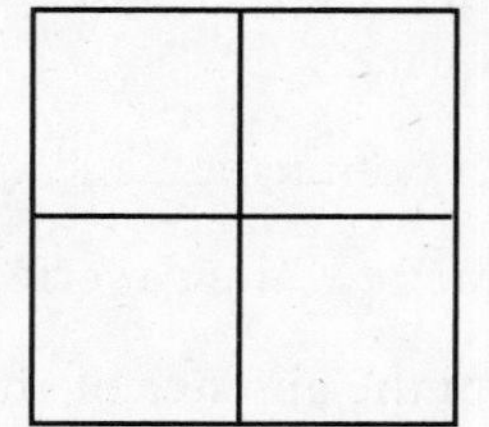

phenotype: ____________________

Chapter 12 **Patterns of Heredity and Human Genetics,** *continued*

Reinforcement and Study Guide

Section 12.2 When Heredity Follows Different Rules, continued

For each statement below, write true or false.

______________ **6.** Traits controlled by more than two alleles are said to have multiple alleles.

______________ **7.** Multiple alleles can be studied only in individuals.

______________ **8.** In humans, there are 23 pairs of matching homologous chromosomes called autosomes.

______________ **9.** Two chromosomes called the sex chromosomes determine the sex of an individual.

______________ **10.** The sex chromosomes of a human male are XX, and the sex chromosomes of a human female are XY.

______________ **11.** Traits controlled by genes located on sex chromosomes are called sex-linked traits.

______________ **12.** The first known example of sex-linked inheritance was discovered in pea plants.

In your textbook, read about environmental influences.

Answer the following questions.

13. What characteristics of an organism can affect gene function?

____

____

14. Do the internal environments of males and females differ? Explain.

____

____

15. What are some environmental factors that can influence gene expression?

____

____

16. Give two examples of how an environmental factor can affect the expression of a phenotype.

____

____

Chapter 12 **Patterns of Heredity and Human Genetics,** *continued*

Reinforcement and Study Guide

Section 12.3 Complex Inheritance of Human Traits

In your textbook, read about multiple alleles in humans.

Complete the table by filling in the missing information.

Genotypes	Human Blood Groups Surface Molecules	Phenotypes
1.	A	
2.		B
3.	A and B	AB
4.	none	

Complete each statement.

5. Blood groups are a classic example of ________________ inheritance in humans.

6. The alleles ________________ are always both expressed.

7. The alleles I^A and I^B are ________________, meaning they are always both expressed.

8. I^A and I^B are dominant to ________________.

9. Blood typing is necessary before a person can receive a ________________.

10. A child who inherits I^A from his mother and I^B from his father will have type ________________ blood.

11. A child whose parents both have type O blood will have type ________________ blood.

12. If a woman with blood type A has a baby with blood type AB, a man with blood type O ________________ be the father.

Chapter 13 **Genetic Technology**

Reinforcement and Study Guide

Section 13.1 Applied Genetics

In your textbook, read about selective breeding and determining genotypes.

Complete each statement.

1. Organisms that are homozygous dominant and those that are ________________ for a trait controlled by Mendelian inheritance have the same phenotype.

2. A ________________ determines whether an organism is heterozygous or homozygous dominant for a trait.

3. Usually the parent with the known genotype is ________________________ for the trait in question.

4. When two cultivars are crossed, their offspring will be ________________ .

Answer the following.

5. A breeder performs a testcross to determine whether an Alaskan malamute is homozygous dominant (*DD*) or heterozygous (*Dd*) for a recessive dwarf allele. Half the offspring appear dwarf. What is the genotype of the unknown dog? Complete the Punnett square to verify your answer.

____

6. What results would be expected if the unknown dog was homozygous dominant (*DD*)? Complete the Punnett square to verify your answer.

____

____

Section 13.2 Recombinant DNA Technology

In your textbook, read about gene engineering.

For each item in Column A, write the letter of the matching item in Column B.

Column A

________ **1.** Bacterial proteins that have the ability to cut both strands of the DNA molecule at certain points

________ **2.** Contain foreign DNA

________ **3.** Is made by connecting segments of DNA from different sources

________ **4.** General term for a vehicle used to transfer a foreign DNA fragment into a host cell

________ **5.** A small ring of DNA found in a bacterial cell

________ **6.** The procedure for cleaving DNA from an organism into small segments, and inserting the segments into another organism

Column B

a. recombinant DNA

b. vector

c. restriction enzymes

d. plasmid

e. transgenic organisms

f. genetic engineering or recombinant DNA technology

Complete the table by checking the correct column for each vector.

Vectors	Mechanical	Biological
7. Viruses		
8. Micropipette		
9. Metal bullets		
10. Plasmids		

In your textbook, read about applications of DNA technology.

Complete the table by checking the correct column for each statement.

Statement	Bacteria	Transgenic Plant(s)	Animal(s)
11. Employed in the production of growth hormone to treat dwarfism and insulin to treat diabetes			
12. Difficult to produce because of thick cell walls and few biological vectors			
13. Some can be made using a bacterium that normally causes tumor-like galls.			
14. Contain many genes common to humans			
15. Have been engineered to break down pollutants into harmless products			
16. The first patented organism			
17. Produced using mechanical vectors such as the gene gun			
18. Produce phenylalanine, an amino acid needed to make artificial sweeteners			
19. In the future, they will be more nutritious and be able to grow in unfavorable conditions.			
20. Helps scientists to learn about human diseases			
21. Produce insulin, a hormone used in treating diabetes			
22. Produced by using a micropipette to inject DNA into unfertilized eggs			
23. Contain foreign genes that slow down the process of spoilage			

Genetic Technology, *continued*

Section 13.3 The Human Genome

In your textbook, read about mapping and sequencing the human genome and applications of the Human Genome Project.

Determine if the statement is true. If it is not, rewrite the italicized part to make it true.

1. The human genome consists of approximately *1000* genes located on 46 chromosomes.

2. Scientists *have* determined the exact chromosomal location of all genes.

3. The genetic map that shows the location of genes on a chromosome is called a *pedigree map*.

4. Instead of examining actual offspring, scientists examine *egg* cells to create linkage maps.

5. Gene therapy is being performed on patients suffering from *sickle-cell anemia*.

6. *Electrolysis* can be used to separate DNA fragments.

Answer the following questions.

7. What is the Human Genome Project?

8. Why is mapping by linkage data extremely inefficient in humans?

9. What are the three areas of current research that utilize chromosome maps?

10. Why is DNA fingerprinting reliable?

Name Date Class

BioDigest 4 **Reinforcement and Study Guide**

Genetics

In your textbook, read about simple Mendelian inheritance and meiosis.

Complete each statement.

1. A trait is ______________ if only one allele is needed for that trait to be expressed. If both alleles are needed for the trait to be expressed, the trait is ______________ .

2. When a *TT* tall pea plant is crossed with a *tt* short pea plant, there is a 100% probability that all offspring will be ______________ and have the genotype ______________ .

3. Unlike mitosis, ______________ produces cells that contain only one copy of each ______________ .

4. ______________ and the rearrangement of alleles during ______________ provide mechanisms for genetic variability.

In your textbook, read about producing physical traits and complex inheritance patterns.

Predict the outcome of the following crosses. Use Punnett squares to support your answers.

5. Homozygous short × Homozygous short

____

6. Heterozygous for purple flowers × Heterozygous for purple flowers

____

____

____

7. Heterozygous pink snapdragon × Heterozygous pink snapdragon

____

____

____

____

Sequence the steps in protein synthesis from 1 to 4.

______ **8.** Amino acids bond together to form a protein.

______ **9.** Sequence of bases in DNA is copied into mRNA.

______ **10.** tRNA molecules bring appropriate amino acids to the mRNA on the ribosome.

______ **11.** mRNA leaves the cell nucleus.

For each item in Column A, write the letter of the matching item in Column B.

	Column A	Column B
______	**12.** Results in an mRNA copy of DNA	**a.** incomplete dominance
______	**13.** Sequence of three bases in mRNA	**b.** X-linked trait
______	**14.** Site of translation	**c.** ribosome
______	**15.** Governed by several genes	**d.** transcription
______	**16.** Heterozygote has an intermediate phenotype.	**e.** translation
______	**17.** Double-stranded molecule that stores and transmits genetic information	**f.** codon
______	**18.** More likely to appear in males than in females	**g.** polygenic inheritance
______	**19.** Results in a sequence of amino acids	**h.** DNA

In your textbook, read about recombinant DNA technology and gene therapy.

Sequence the steps to making recombinant DNA from 1 to 5.

______ **20.** The plasmid becomes part of a host cell's chromosome.

______ **21.** A DNA fragment is inserted into a plasmid.

______ **22.** The DNA fragment replicates during cell division.

______ **23.** The plasmid enters a host bacterial cell.

______ **24.** A host cell produces a protein that it would not have produced naturally.

Answer the following questions.

25. What is gene therapy?

______

______

26. What are clones?

______

27. What is a vector? Give two examples of vectors.

______

______

______

Chapter 14

The History of Life

Reinforcement and Study Guide

Section 14.1 The Record of Life

In your textbook, read about the early history of Earth.

For each statement below, write true or false.

______ **1.** Earth is thought to have formed about 4.6 billion years ago.

______ **2.** The conditions on primitive Earth were very suitable for life.

______ **3.** Geological events on Earth set up conditions that would play a major role in the evolution of life on Earth.

______ **4.** Violent rainstorms beginning 3.9 million years ago formed Earth's oceans.

______ **5.** The first organisms appeared on land between 3.9 and 3.5 billion years ago.

In your textbook, read about a history in the rocks.

For each statement in Column A, write the letter of the matching item in Column B.

	Column A	Column B
______	**6.** A footprint, trail, or burrow, providing evidence of animal activity	**a.** petrified fossil
______	**7.** A fossil embedded in tree sap, valuable because the organism is preserved intact	**b.** imprint
______	**8.** An exact stone copy of an original organism, the hard parts of which have been penetrated and replaced by minerals	**c.** trace fossil
______	**9.** Any evidence of an organism that lived long ago	**d.** cast
______	**10.** The fossil of a thin object, such as a leaf or feather, that falls into sediments and leaves an outline when the sediments hardened	**e.** amber-preserved
______	**11.** An empty space left in rock, showing the exact shape of the organism that was buried and decayed there	**f.** fossil
______	**12.** An object formed when a mold is filled in by minerals from the surrounding rock	**g.** mold

Chapter 14

The History of Life, *continued*

Reinforcement and Study Guide

Section 14.1 The Record of Life, continued

In your textbook, read about the age of a fossil.

Answer the following questions.

13. Explain how relative dating works.

14. What is the limitation of relative dating?

15. What dating technique is often used by paleontologists to determine the specific age of a fossil?

16. How do scientists use this dating technique to determine the ages of rocks or fossils?

In your textbook, read about a trip through geologic time.

Complete the table by checking the correct column for each statement.

Statement	Era			
	Pre-Cambrian	**Paleozoic**	**Mesozoic**	**Cenozoic**
17. The first photosynthetic bacteria form dome-shaped structures called stromatolites.				
18. Primates evolve and diversify.				
19. Divided into three periods: Triassic, Jurassic, and Cretaceous				
20. An explosion of life, characterized by the appearance of many types of invertebrates and plant phyla				
21. Mammals appear.				
22. Dinosaurs roam Earth, and the ancestors of modern birds evolve.				
23. Flowering plants appear.				
24. Amphibians and reptiles appear.				

Section 14.2 The Origin of Life

In your textbook, read about origins: the early ideas.

Use each of the terms below just once to complete the passage.

microorganisms	**vital force**	**Louis Pasteur**	**biogenesis**
nonliving matter	**S-shaped**	**disproved**	**Francesco Redi**
organisms	**broth**	**microscope**	**spontaneous generation**
spontaneously	**air**		

Early scientists believed that life arose from **(1)** ____________________ through a process they called **(2)** ____________________. In 1668, the Italian physician **(3)** ____________________ conducted an experiment with flies that **(4)** ____________________ this idea. At about the same time, biologists began to use an important new research tool, the **(5)** ____________________. They soon discovered the vast world of **(6)** ____________________. The number and diversity of these organisms was so great that scientists were led to believe once again that these organisms must have arisen **(7)** ____________________. By the mid-1800s, however, **(8)** ____________________ was able to disprove this hypothesis once and for all. He set up an experiment, using flasks with unique **(9)** ____________________ necks. These flasks allowed **(10)** ____________________, but no organisms, to come into contact with a broth containing nutrients. If some **(11)** ____________________ existed, as had been suggested, it would be able to get into the **(12)** ____________________ through the open neck of the flask. His experiment proved that organisms arise only from other **(13)** ____________________. This idea, called **(14)** ____________________, is one of the cornerstones of biology today.

Determine if the statement is true. If it is not, rewrite the italicized part to make it true.

15. Biogenesis *explains* how life began on Earth.

____

16. For life to begin, simple *inorganic* molecules had to be formed and then organized into complex molecules.

____

17. Several billion years ago, Earth's atmosphere had no free *methane*.

____

Section 14.2 The Origin of Life, continued

18. Primitive Earth's atmosphere may have been composed of water vapor, hydrogen, methane, and *ammonia*. ______________________________

19. In the early 1900s, Alexander Oparin proposed a widely accepted hypothesis that life began *on land*.

____

20. *Pasteur* hypothesized that many chemical reactions occurring in the atmosphere resulted in the formation of a primordial soup. ______________________________

21. In 1953, Miller and Urey tested Oparin's hypothesis by simulating the conditions of *modern* Earth in the laboratory. ______________________________

22. Miller and Urey showed that organic compounds, including *nucleic acids* and sugars, could be formed in the laboratory, just as had been predicted. ______________________________

23. This "life-in-a-test-tube" experiment of Miller and Urey provides support for some modern hypotheses of *biogenesis*. ______________________________

24. Sidney Fox took Miller and Urey's experiment further and showed how amino acids could cluster to form *protocells*. ______________________________

In your textbook, read about the evolution of cells.

Answer the following questions.

25. Describe the likely characteristics of the first organisms on Earth.

____

____

____

26. What is an autotroph? What factors helped them thrive on Earth?

____

____

____

27. What present-day organisms may be similar to the first autotrophs? Why?

____

28. What change occurred in Earth's atmosphere after the evolution of photosynthesizing prokaryotes? Why?

____

____

Name Date Class

Chapter 15 **The Theory of Evolution**

Reinforcement and Study Guide

Section 15.1 Natural Selection and the Evidence for Evolution

In your textbook, read about Charles Darwin and natural selection.

For each statement, write true or false.

______ **1.** H.M.S. *Beagle*, upon which Charles Darwin served as naturalist, set sail on a collecting and mapping expedition in 1831.

______ **2.** The environments that Darwin studied exhibited little biological diversity.

______ **3.** By careful anatomical study, Darwin found that the many species of plants and animals on the Galapagos Islands were unique and bore no relation to species seen in other parts of the world.

______ **4.** The tortoises of the Galapagos Islands are among the largest on Earth.

______ **5.** After returning to England, Darwin studied his collections for 10 years.

______ **6.** Darwin named the process by which evolution proceeds *artificial selection*.

You are a naturalist who traveled to the Galapagos Islands. Below are excerpts from field notes. Next to each set of notes, write a heading. Use these choices: Overproduction of Offspring, Natural Selection, Struggle for Existence, Variation.

7.

Field Notes

Female finches found on the Galapagos Islands lay enormous numbers of eggs.

______

8.

Field Notes

These finches compete for a particular species of insect that inhabits the small holes found in tree bark.

______

9.

Field Notes

Some finches' beaks are long, some are short.The finches with long beaks are better adapted to remove the insects from the bark.

______

10.

Field Notes

The finches with the long beaks survive and produce greater numbers of offspring with long beaks.

______

Chapter 15 **The Theory of Evolution,** *continued*

Reinforcement and Study Guide

Section 15.1 Natural Selection and the Evidence for Evolution, continued

In your textbook, read about natural selection and adaptations.

Identify the type of structural adaptation that the statement describes. If the statement applies to both, write both. Use these choices: mimicry, camouflage, both.

______________ **11.** Enable(s) an organism to blend in with its surroundings

______________ **12.** Provide(s) protection for an organism by copying the appearance of another species

______________ **13.** The coloration of a flounder that allows the fish to avoid predators

______________ **14.** Involve(s) changes to the external appearance of an organism

______________ **15.** A flower that looks like a female bee

In your textbook, read about evidence for evolution.

Complete the chart by checking the kind of evidence described.

Evidence	Type of Evidence				
	Homologous Structure	**Analogous Structure**	**Vestigial Structure**	**Embryological Development**	**Genetic Comparisons**
16. A modified structure seen among different groups of descendants					
17. In the earliest stages of development, a tail and gill slits can be seen in fish, birds, rabbits, and mammals.					
18. Exemplified by forelimbs of bats, penguins, lizards, and monkeys					
19. The forelimbs of flightless birds					
20. DNA and RNA comparisons may lead to evolutionary trees.					
21. Bird and butterfly wings have same function but different structures					
22. A body structure reduced in function but may have been used in an ancestor					

Section 15.2 Mechanisms of Evolution

In your textbook, read about population genetics and evolution.

Determine if the statement is true. If it is not, rewrite the italicized part to make it true.

1. *Adaptations* of species are determined by the genes contained in the DNA code. ________________

2. When Charles *Mendel* developed the theory of natural selection in the 1800s, he did not include a genetic explanation. ________________

3. Natural selection can act upon an individual's *genotype*, the external expression of genes. ________________

4. Natural selection operates on *an individual* over many generations. ________________

5. The entire collection of genes among a population is its *gene frequency*. ________________

6. If you know the *phenotypes* of all the organisms in a population, you can calculate the allelic frequency of the population. ________________

7. A population in which frequency of alleles *changes* from generation to generation is said to be in genetic equilibrium. ________________

8. A population that is in *genetic equilibrium* is not evolving. ________________

9. Any factor that affects *phenotype* can change allelic frequencies, thereby disrupting the genetic equilibrium of populations. ________________

10. Many *migrations* are caused by factors in the environment, such as radiation or chemicals, but others happen by chance. ________________

11. Mutations are *important* in evolution because they result in genetic changes in the gene pool. ________________

12. Genetic *equilibrium* is the alteration of allelic frequencies by chance processes. ________________

13. Genetic drift is more likely to occur in *large* populations. ________________

14. The factor that causes the greatest change in gene pools is *mutation*. ________________

15. The type of natural selection by which one of the extreme forms of a trait is favored is called *disruptive selection*. ________________

Chapter 15 **The Theory of Evolution,** *continued*

Reinforcement and Study Guide

Section 15.2 Mechanisms of Evolution, continued

In your textbook, read about the evolution of species.

Complete each statement.

16. ____________________ can occur only when either interbreeding or the production of fertile offspring is prevented among members of a population.

17. _________________________ occurs when formerly interbreeding organisms are prevented from producing fertile offspring.

18. Polyploid speciation is perhaps the fastest form of speciation because it results in immediate _________________________ .

19. The hypothesis that species originate through a slow buildup of new adaptations is known as ____________________ .

20. This hypothesis is supported by evidence from the ____________________ record.

21. The hypothesis of __________________________ states that speciation may occur rapidly.

In your textbook, read about patterns of evolution.

Answer the following questions.

22. What happened to the ancestor of the honey creeper when it left the mainland and encountered the diverse niches of Hawaii?

____

____

23. What is adaptive radiation?

____

____

24. Adaptive radiation is one example of divergent evolution. When does divergent evolution occur?

____

____

____

25. When will convergent evolution occur?

____

____

Name Date Class

Chapter 16 Primate Evolution

Reinforcement and Study Guide

Section 16.1 Primate Adaptation and Evolution

In your textbook, read about the characteristics of a primate.

Complete the chart by checking those structures or functions that are characteristic of primates.

Structure/Function	Primate
1. Round head	
2. Flattened face	
3. Small head	
4. Large relative brain size	
5. Highly developed vision	
6. Poor vision	
7. Binocular vision	
8. Color vision	
9. Color-blind	
10. Vision the dominant sense	
11. Smell the dominant sense	
12. Immobile joints	
13. Flexible shoulder joints	
14. Skeleton adapted for movement among trees	
15. Skeleton adapted for swimming	
16. Hands and feet equipped with claws	
17. Hands and feet equipped with nails	
18. Eyes face to the side	
19. Feet constructed for grasping	
20. Opposable thumbs	

In your textbook, read about primate origins.

For each statement below, write true or false.

______________ **21.** Scientists believe that primates evolved about 66,000 years ago.

______________ **22.** The earliest primate may have been a prosimianlike animal called *Purgatorius.*

______________ **23.** Anthropoids are a group of small-bodied primates.

______________ **24.** Prosimians include lemurs and tarsiers.

______________ **25.** Prosimians can be found in the tropical forests of South America.

Section 16.1 Primate Adaptation and Evolution, continued

Identify the following pictures. Use these choices: baboon, tarsier, spider monkey. Then on the second line write the group that is represented by the picture. Use these choices: New World monkey, Old World monkey, prosimian.

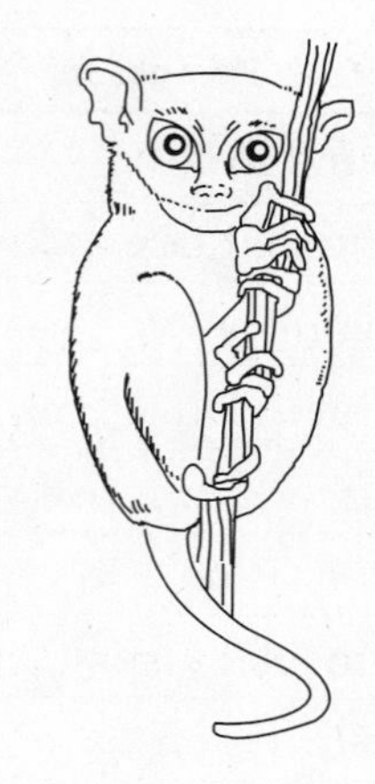

26. ______________________

____

27. ______________________

____

28. ______________________

____

Answer the following questions.

29. What do similarities among monkeys, apes, and humans indicate about their evolution?

____

____

30. According to the fossil record, what were the first modern anthropoids to evolve and about when did they evolve?

____

____

31. What is the evolutionary history of primates based on?

____

____

32. What may have led to the eventual speciation of baboons and other ground-living monkeys?

____

____

33. What does DNA analysis of modern hominoids suggest about their evolutionary history?

____

____

Name Date Class

In your textbook, read about hominids.

Answer the following questions.

1. What is an australopithecine? ______
2. What fossil skull did Raymond Dart discover in Africa in 1924? ______
3. Why was *A. africanus* unlike any primate fossil skull that Dart had ever seen? ______
4. What did the position of the foramen magnum indicate to Dart? ______

Label the following skulls. Use these choices: chimpanzee, human, *A. afarensis*

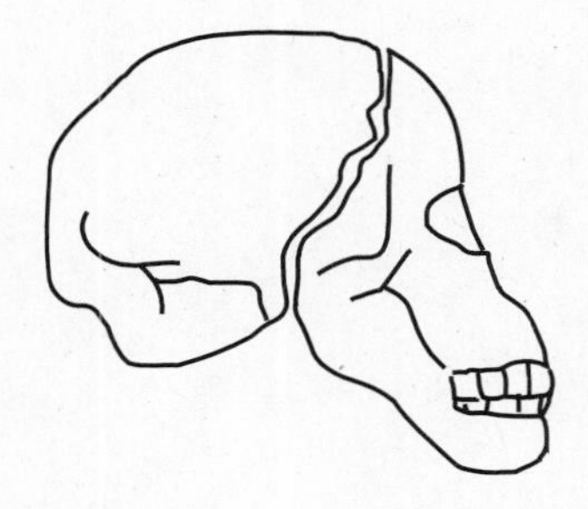

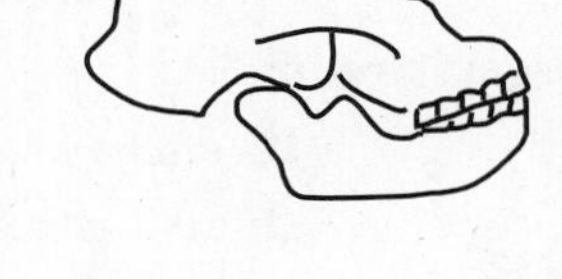

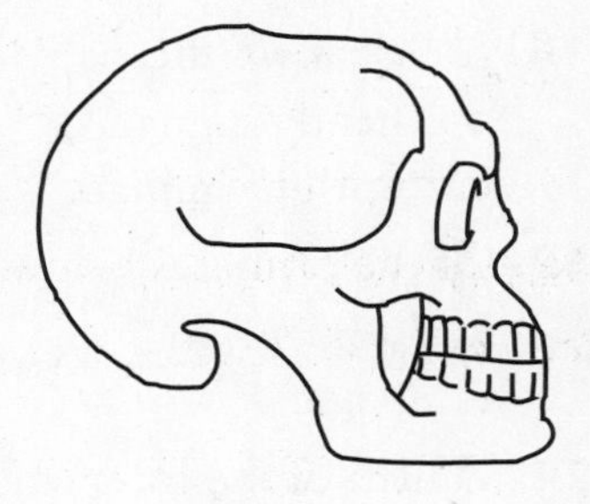

5. ______ 6. ______ 7. ______

For each statement below, write <u>true</u> or <u>false</u>.

______ 8. Much of what scientists know about australopithecines comes from the "Lucy"skeleton.

______ 9. "Lucy" is 3.5 billion years old.

______ 10. "Lucy" is classified as *A. africanus.*

______ 11. *A. afarensis* is the earliest known hominid species.

______ 12. *A. afarensis* walked on all four legs and had a humanlike brain.

______ 13. Australopithecines are alive today and can be found in southern Africa and Asia.

______ 14. Australopithecines probably played a role in the evolution of modern hominids.

Chapter 16 **Primate Evolution,** *continued*

Reinforcement and Study Guide

Section 16.2 Human Ancestry, continued

In your textbook, read about the emergence of modern humans.

Circle the letter of the choice that best completes the statement or answers the question.

15. The first skull of *Homo habilis* was discovered by
- **a.** Raymond Dart.
- **b.** Louis and Mary Leakey.
- **c.** Donald Johanson.
- **d.** Gert Terblance.

16. When compared to an australopithecine skull, the *Homo habilis* skull is
- **a.** more humanlike.
- **b.** less humanlike.
- **c.** more apelike.
- **d.** exactly the same.

17. Which of the following is *not* true about *Homo habilis?*
- **a.** They existed between 1.5 and 2 million years ago.
- **b.** They were the first hominids to make and use tools.
- **c.** They were probably scavengers of their food.
- **d.** They gave rise to *A. africanus*.

18. *Homo habilis* means
- **a.** "handy human."
- **b.** "tool-using human."
- **c.** "upright human."
- **d.** "talking human."

19. Of the primates below, which has the largest brain?
- **a.** *Homo habilis*
- **b.** *Homo erectus*
- **c.** an ape
- **d.** an australopithecine

20. Which of the following is *not* true about *Homo erectus*?
- **a.** They probably hunted.
- **b.** They were the first hominids to use fire.
- **c.** They may have given rise to hominids that resemble modern humans.
- **d.** They were found only in Africa.

21. *Homo sapiens* includes
- **a.** Neanderthals.
- **b.** australopithecines.
- **c.** *A. africanus.*
- **d.** *A. afarensis.*

Determine whether each statement below best describes Neanderthals, Cro-Magnons, or both.

____________________ **22.** They lived in caves during the ice ages.

____________________ **23.** They are identical to modern humans in height, skull, and teeth structure.

____________________ **24.** They may have been the first hominids to develop religious views.

____________________ **25.** They may have used language.

____________________ **26.** They were talented toolmakers and artists.

Name ____________________ Date __________ Class __________

Chapter 17 **Organizing Life's Diversity**

Reinforcement and Study Guide

Section 17.1 Classification

In your textbook, read about how classification began and about biological classification.

For each item in Column A, write the letter of the matching item in Column B.

	Column A	Column B
______	**1.** Grouping objects or information based on similarities	**a.** Aristotle
______	**2.** Naming system that gives each organism a two-word name	**b.** Linnaeus
______	**3.** Developed the first system of classification	**c.** genus
______	**4.** Branch of biology that groups and names organisms	**d.** classification
______	**5.** Designed a system of classifying organisms based on their physical and structural similarities	**e.** taxonomy
______	**6.** Consists of a group of similar species	**f.** binomial nomenclature

Determine if the statement is true. If it is not, rewrite the italicized part to make it true.

7. The scientific name of a species consists of a *family* name and a descriptive name.

____

8. The scientific name of modern humans is *Homo sapiens*.

____

9. *Latin* is the language of scientific names.

____

10. The *scientific* names of organisms can be misleading.

____

11. Taxonomists try to identify the *evolutionary relationships* among organisms.

____

12. Besides comparing the structures of organisms, taxonomists also compare the organisms' geographic distribution and *chemical makeup*.

____

13. Similarities between living species and extinct species *cannot* be used to determine their relationship to each other.

____

14. Because the bones of some dinosaurs have large internal spaces, some scientists think dinosaurs are more closely related to *amphibians* than to reptiles.

____

15. Classification can be useful in identifying the *characteristics* of an unknown organism.

____

Section 17.1 Classification, continued

In your textbook, read about how living things are classified.

Examine the table showing the classification of four organisms. Then answer the questions.

Taxon	Green Frog	Mountain Lion	Domestic Dog	Human
Kingdom	Animalia	Animalia	Animalia	Animalia
Phylum	Chordata	Chordata	Chordata	Chordata
Class	Amphibia	Mammalia	Mammalia	Mammalia
Order	Anura	Carnivora	Carnivora	Primates
Family	Ranidae	Felidae	Canidae	Hominidae
Genus	*Rana*	*Felis*	*Canis*	*Homo*
Species	*Rana clamitans*	*Felis concolor*	*Canis familiaris*	*Homo sapiens*

16. Which taxon includes the most specific characteristics? ______

17. Which taxon includes the broadest characteristics? ______

18. Which taxon includes more species, an order or a family? ______

19. Which taxon includes only organisms that can successfully interbreed? ______

20. If two organisms belong to the same family, what other taxonomic groups do the organisms have in common.

______

21. Which two organisms in the chart are most closely related? Explain.

______

______

______

______

22. To which taxa do all four organisms belong?

______

23. Which class does not include animals that have hair or fur? ______

24. What is the order, family, and genus of a human?

______

25. Using the information in the chart, what can you conclude about the classification taxa of an organism with the scientific name *Rana temporaria?*

______

______

______

In your textbook, read about how evolutionary relationships are determined.

Explain how scientists use each item below to determine the evolutionary relationships among organisms.

1. structural similarities: ______________________________

____

2. breeding behavior: ______________________________

____

3. geographical distribution: ______________________________

____

4. chromosome comparisons: ______________________________

____

5. biochemistry: ______________________________

____

In your textbook, read about phylogenetic classification: models.

Use the cladogram to answer the questions.

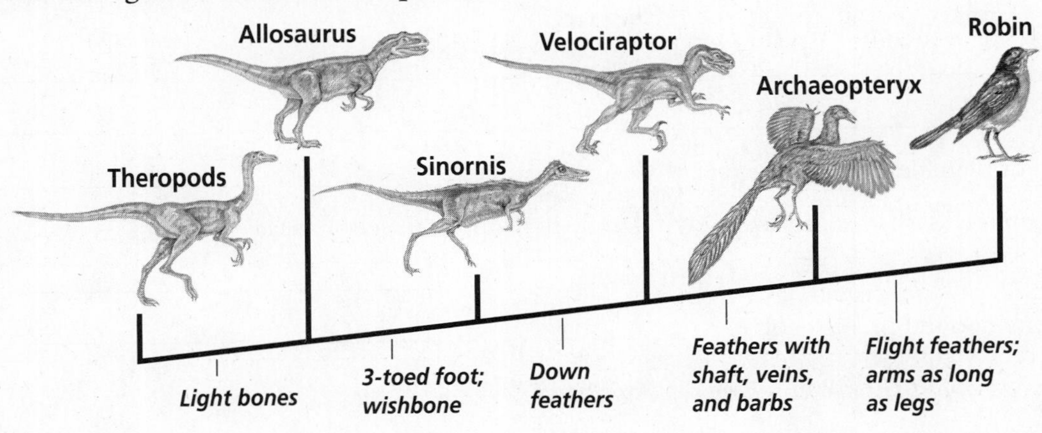

6. What five probable ancestors of the modern bird (robin) are shown on the cladogram?

____

7. Which dinosaur is probably the most recent common ancestor of *Velociraptor* and *Archaeopteryx*?

____

8. Which traits shown on the cladogram are shared by *Archaeopteryx* and modern birds?

____

____

Section 17.2 The Six Kingdoms, *continued*

Use the fanlike phylogenetic diagram to answer the questions.

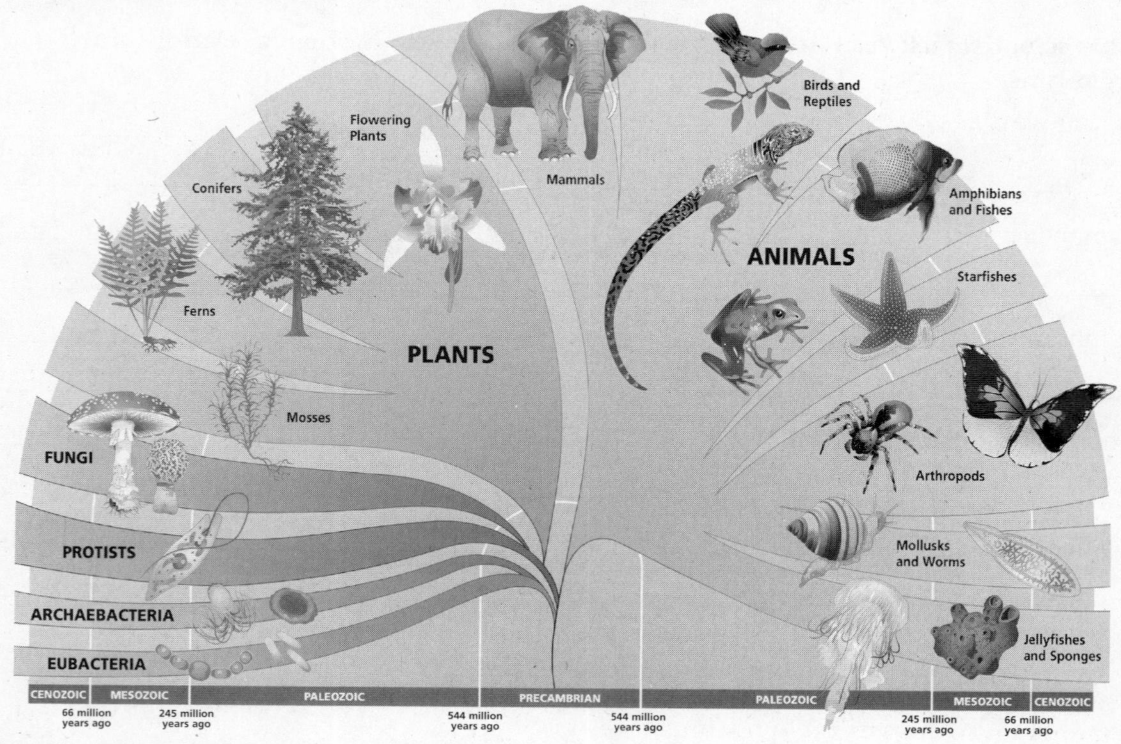

9. How does the fanlike diagram differ from a cladogram?

____

____

10. Which group of plants evolved most recently? ______________________

11. To which group are starfishes more closely related, arthropods or jellyfishes? ____________

12. Which group of animals includes the fewest species? ______________________

13. About how long ago did plants evolve? ______________________

In your textbook, read about the six kingdoms of organisms.

Circle the letter of the choice that best completes the statement or answers the question.

14. Organisms that do not have a nucleus bounded by a membrane are

a. multicellular. **b.** eukaryotes. **c.** protists. **d.** prokaryotes.

15. Fungi obtain food by

a. photosynthesis. **b.** chemosynthesis.

c. endocytosis. **d.** absorbing nutrients from organic materials.

16. Animals are

a. autotrophs. **b.** heterotrophs. **c.** prokaryotes. **d.** stationary.

Name Date Class

BioDigest 5

Reinforcement and Study Guide

Change Through Time

In your textbook, read about the Geologic Time Scale.

Complete the table.

Era	Time period	Biologic event
1.	4.6 billion–600 million years ago	**2.**
3.	600 million–245 million years ago	**4.**
5.	245 million–66 million years ago	**6.**
7.	66 million years ago–present	**8.**

In your textbook, read about origin of life theories.

Complete each statement.

9. Spontaneous generation assumes that life arises spontaneously from ______________________ .

10. Francesco Redi and Louis Pasteur designed ______________________ to disprove spontaneous generation.

11. The theory of ______________________ states that life comes only from pre-existing life.

12. Clusters of organic molecules might have formed ______________________, which may have evolved into the first true cells.

Order the evolutionary development of the following organisms from 1 to 4.

__________ **13.** chemosynthetic prokaryotes

__________ **14.** eukaryotes

__________ **15.** heterotrophic prokaryotes

__________ **16.** oxygen-producing photosynthetic prokaryotes

In your textbook, read about the evidence and mechanics of evolution.

Answer the following questions.

17. What assumption is made in the relative dating of fossils? ______________________

____

18. What are homologous structures? ______________________

____

BioDigest 5

Change Through Time, *continued*

Reinforcement and Study Guide

For each statement below, write true or false.

________ **19.** Evolution occurs when a population's genetic equilibrium remains unchanged.

________ **20.** Mutations, genetic drift, and migration may disrupt the genetic equilibrium of populations.

________ **21.** Stabilizing selection favors the survival of a population's average individuals for a feature.

________ **22.** Disruptive selection occurs when an extreme feature is naturally selected.

________ **23.** Adaptive radiation occurs when species that once were similar to an ancestral species become increasingly distinct due to natural selection pressures.

In your textbook, read about primate evolution.

For each item in Column A, write the letter of the matching item in Column B.

Column A

________ **24.** Primate adaptation

________ **25.** Primate category that includes humans and apes

________ **26.** Characteristic of New World monkeys

________ **27.** Appearing in fossil record about 2 million years ago along with stone tools

________ **28.** Possible human ancestors dating from 5 to 8 million years ago

Column B

a. anthropoids

b. Australopithecines

c. genus *Homo*

d. opposable thumb

e. prehensile tail

In your textbook, read about organizing life's diversity.

Look at the taxonomic classification of a bobcat shown below. Answer the questions.

29. What is the largest taxon in this classification system?

____

30. What is the scientific name of a bobcat?

____

Taxon	Name
Kingdom	*Animalia*
Phylum	*Chordata*
Class	*Mammalia*
Order	*Carnivora*
Family	*Felidae*
Genus	*Lynx*
Species	*rufus*

Chapter 18

Viruses and Bacteria

Reinforcement and Study Guide

Section 18.1 Viruses

In your textbook, read about the characteristics of a virus.

For each item in Column A, write the letter of the matching item in Column B.

	Column A	Column B
______	**1.** Genetic material of a virus	**a.** virus
______	**2.** Where a virus attaches to a host cell	**b.** T4 phage
______	**3.** Nonliving particle that replicates inside a living cell	**c.** DNA or RNA
______	**4.** A virus's protein coat	**d.** capsid
______	**5.** Interlocks with a molecular shape in a host cell's plasma membrane	**e.** receptor site
______	**6.** Layer that surrounds the capsid of some viruses	**f.** envelope
______	**7.** A virus that infects *E. coli* bacteria	**g.** host
______	**8.** A cell in which a virus replicates	**h.** attachment protein

In your textbook, read about viral replication cycles.

Complete the table by checking the correct column for each statement.

Statement	Lytic Cycle	Lysogenic Cycle
9. Viral genes are expressed immediately after the virus infects the host cell.		
10. Many new viruses are assembled.		
11. This cycle is preceded by a virus entering a host cell.		
12. Viral DNA is integrated into the host cell's chromosome.		
13. Viruses are released from the host cell by lysis or exocytosis.		
14. Reverse transcriptase is used to make DNA from the RNA of a retrovirus.		
15. A provirus is replicated along with the host cell's chromosome.		

Section 18.1 Viruses, continued

Use each of the terms below just once to complete the passage.

DNA	**white blood cells**	**lysogenic**
lytic	**AIDS**	**proviruses**

Many disease-causing viruses have both lytic and **(16)** ________________ cycles. For example, when HIVs infect **(17)** ________________, the viruses enter a lysogenic cycle. Their genetic material becomes incorporated into the **(18)** ________________ of the white blood cells, forming **(19)** ________________. When this happens, the white blood cells still function normally, and the person may not appear ill. Eventually, the proviruses enter a **(20)** ________________ cycle, killing the white blood cells. As a result, the person loses the ability to fight diseases and develops **(21)** ________________.

In your textbook, read about viruses and cancer, plant viruses, and the origin of viruses.

Determine if the statement is true. If it is not, rewrite the italicized part to make it true.

________________ **22.** Some viruses can change normal cells to *tumor* cells.

________________ **23.** Retroviruses and the papilloma virus, which causes *hepatitis B*, are examples of tumor viruses.

________________ **24.** *All* plant viruses cause diseases in plants.

________________ **25.** The first virus ever identified was the plant virus called *tobacco mosaic virus*.

________________ **26.** The patterns of color in some flowers are caused by *tumor* viruses.

________________ **27.** Tumor viruses contain genes that are found in *normal* cells.

________________ **28.** Scientists think viruses originated from *their host cells*.

Name Date Class

Section 18.2 Archaebacteria and Eubacteria

In your textbook, read about the diversity of prokaryotes and about the characteristics of bacteria.

Answer the following questions.

1. What are three types of environments in which archaebacteria are found? ______

2. In what three ways do eubacteria obtain nutrients? ______

3. How does a bacterium's cell wall protect it? ______

4. Where is the genetic material of a bacterium found? ______

5. What structure do some bacteria use to move? ______

6. What is the difference between gram-positive bacteria and gram-negative bacteria? ______

7. What are three different shapes of bacteria? ______

8. Describe the three growth patterns of bacteria and state the prefix used to identify each growth pattern. ______

Identify the type of bacterial reproduction described. Use these choices: binary fission, conjugation.

______ 9. Bacterium with a new genetic makeup is produced.

______ 10. Circular chromosome is copied.

______ 11. Genetic material is transferred through a pilus.

______ 12. Two identical cells are produced.

______ 13. Sexual reproduction occurs.

Chapter 18 **Viruses and Bacteria,** *continued*

Reinforcement and Study Guide

Section 18.2 Archaebacteria and Eubacteria, continued

In your textbook, read about adaptations in bacteria and the importance of bacteria.

Circle the letter of the choice that best completes the statement.

14. Scientists think the first bacteria on Earth were
a. aerobic. **b.** anaerobic. **c.** fatal. **d.** oxygen-dependent.

15. Bacteria that are obligate anaerobes release energy from food by
a. cellular respiration. **b.** using oxygen.
c. using nitrogen. **d.** fermentation.

16. As an endospore, a bacterium
a. produces toxins. **b.** dries out. **c.** causes diseases. **d.** is protected.

17. Botulism is caused by endospores of *C. botulinum* that have
a. been killed. **b.** produced toxins.
c. germinated. **d.** reproduced.

18. Nitrogen is important because all organisms need it to make
a. proteins. **b.** ATP. **c.** DNA. **d.** all of these.

19. The process by which bacteria use enzymes to convert nitrogen gas into ammonia is called
a. nitrogenation. **b.** atmospheric separation.
c. nitrogen fixation. **d.** eutrophication.

20. Bacteria return nutrients to the environment by breaking down
a. dead organic matter. **b.** inorganic materials.
c. enzymes and sugar. **d.** nitrogen in legumes.

21. Bacteria are *not* used to make
a. vinegar. **b.** jams. **c.** cheese. **d.** yogurt.

22. Bacteria are responsible for the following diseases:
a. strep throat and tetanus. **b.** gonorrhea and syphilis.
c. tuberculosis and diphtheria. **d.** all of these.

23. Due to reduced death rates from bacterial diseases and improved sanitation and living conditions, the average person born in the United States today will live to be about
a. 25 years old. **b.** 50 years old.
c. 75 years old. **d.** 90 years old.

Name ____________________ Date __________ Class __________

Chapter 19

Protists

Reinforcement and Study Guide

Section 19.1 The World of Protists

In your textbook, read about what a protist is.

Determine if the statement is true. If it is not, rewrite the italicized part to make it true.

1. The kingdom *Protista* is the most diverse of all six kingdoms. ____________________

2. Protists can be grouped into three general types—*animal-like*, *plantlike*, and *viruslike*.

____

3. All protists are *eukaryotes* that carry on most of their metabolic processes in membrane-bound organelles. ____________________

In your textbook, read about the characteristics and diversity of protozoans.

Complete each statement.

4. The ____________________ -like protists are all unicellular heterotrophs known as protozoans.

5. Amoebas move and change their body shape by forming extensions of their plasma membranes called ____________________ .

6. Amoebas use ____________________ to pump out excess water from their cytoplasm.

7. Most amoebas reproduce by ____________________ in which a parent produces one or more identical offspring by dividing into two cells.

8. One group of protozoans are called ____________________ because they move by whipping one or more flagella from side to side.

9. A paramecium moves by beating thousands of hairlike ____________________ .

10. When food supplies are low, paramecia may reproduce by undergoing a form of ____________________ .

11. Parasitic protozoans called ____________________ live inside their hosts and may reproduce by means of a spore.

12. Malaria is caused by protozoans of the genus ____________________ .

13. The insect that is responsible for transmitting malaria-causing protozoans to humans is the ____________________ .

Chapter 19 **Protists,** *continued*

Reinforcement and Study Guide

Section 19.2 Algae: Plantlike Protists

In your textbook, read about what algae are and about their diversity.

For each item in Column A, write the letter of the item in Column B that completes the statement correctly.

	Column A	Column B
______	**1.** The euglenoids, diatoms, and dinoflagellates are ______ .	**a.** algae
______	**2.** Unicellular protists that are major producers of oxygen in aquatic ecosystems are ______ .	**b.** phyla
______	**3.** Unicellular and multicellular photosynthetic protists are ______ .	**c.** pigments
______	**4.** Most green, red, and brown algae are ______ algae.	**d.** phytoplankton
______	**5.** Photosynthetic ______ are used to classify algae.	**e.** unicellular
______	**6.** Algae are classified into six ______ .	**f.** multicellular

Identify the phylum of the alga shown below and label its parts. Use these choices: flagellum, mitochondrion, pellicle, chloroplast, nucleus, eyespot, contractile vacuole, Euglenophyta.

7. Phylum ______

8. ______

9. ______

10. ______

11. ______

12. ______

13. ______

14. ______

Section 19.2 Algae: Plantlike Protists, continued

Circle the letter of the choice that best completes the statement.

15. When diatoms that have been reproducing asexually reach about one-fourth of their original size, they
- **a.** die.
- **b.** triple in size.
- **c.** reproduce sexually.
- **d.** all of these.

16. Dinoflagellates are unicellular algae that
- **a.** have two flagella.
- **b.** create red tides.
- **c.** have thick cellulose plates.
- **d.** all of these.

17. Red algae are a kind of seaweed having pigments that absorb green, violet, and blue light waves, which allows the algae to
- **a.** live only in fresh water.
- **b.** photosynthesize in limited light.
- **c.** live only in salt water.
- **d.** both a and b.

18. The air bladders of brown algae allow the algae to
- **a.** breathe.
- **b.** reproduce.
- **c.** float near the water's surface.
- **d.** live in salt water.

19. A green alga that forms colonies is
- **a.** Spirogyra.
- **b.** Ulva.
- **c.** Chlamydomonas.
- **d.** Volvox.

In your textbook, read about alternation of generations.

Use each of the terms below just once to complete the passage.

diploid	**gametophyte**	**alternation of generations**	**meiosis**
haploid	**sporophyte**	**spores**	**zygote**

Some algae have a life cycle that has a pattern called **(20)** ______________________________ .
These algae alternate between a(n) **(21)** ___________________________ form that is called the **(22)** ___________________________ because it produces gametes, and a(n) **(23)** ___________________________ form called the **(24)** ___________________________ . When the haploid gametes fuse, they form a(n) **(25)** ___________________________ from which the sporophyte develops. Certain cells in the sporophyte undergo **(26)** ___________________________ to form haploid **(27)** ___________________________ that develop into gametophytes.

Section 19.3 Slime Molds, Water Molds, and Downy Mildews

In your textbook, read about the different kinds of funguslike protists.

Use all the terms in the list below at least once to complete the concept map for funguslike protists.

cell division **three phyla** **spores** **cellular slime molds**
plasmodial slime molds **water molds and mildew** **flagellated reproductive cells**

Funguslike protists

obtain energy by

decomposing organic materials

and can be divided into

1. ______________________

which are

2. ______________________
____

3. ______________________
____

4. ______________________
____

that reproduce by

5. ______________________
____

that reproduce by

6. ______________________
____

that reproduce by

7. ______________________
____

In your textbook, read about the origin of protists.

Answer the following question.

8. What does scientific evidence show is the relationship between protists and other groups of organisms?

____

____

____

____

Chapter 20 **Fungi**

Reinforcement and Study Guide

Section 20.1 What Is a Fungus?

In your textbook, read about the general characteristics of fungi.

Answer the following questions.

1. What are the threadlike filaments in a multicellular fungus called? What do they form?

____

____

____

2. Some hyphae are divided into individual cells by cross walls with pores in them. What are these cross walls called? What purpose do the pores serve?

____

____

____

3. What is one way that fungi are like plants and one way they are unlike plants?

____

____

____

____

In your textbook, read about adaptations in fungi.

Determine if the statement is true. If it is not, rewrite the italicized part to make it true.

4. Many fungi are decomposers, which *break down organic substances* into raw materials that can be used by other organisms. ______________________________

5. Fungi use *cellular digestion* to obtain nutrients. ______________________________

6. Hyphae release *digestive enzymes* that break down molecules in their food source.

____

7. *Mutualistic* fungi are decomposers. ______________________________

8. *Parasitic fungi grow spores* into host cells and absorb the cell's nutrients.

____

Section 20.1 What Is a Fungus?, continued

In your textbook, read about reproduction in fungi.

Complete each statement.

9. Fungi reproduce ______________________ by fragmentation, budding, or producing spores.

10. In ______________________, pieces of hyphae grow into new mycelia.

11. The process of a parent cell undergoing mitosis and producing a new individual that pinches off, matures, and separates from the parent is called ______________________.

12. When environmental conditions are right, a ______________________ may germinate and produce a threadlike ______________________ that will grow into a mycelium.

13. Some hyphae grow away from the mycelium to produce a spore-containing structure called a ______________________.

14. In most fungi, the structures that support ______________________ are the only part of the fungus that can be seen.

15. Fungi may produce spores by ______________________ or ______________________.

16. Many adaptations of fungi for survival involve ______________________.

17. ______________________ protect spores and keep them from from drying out until they are released.

18. A single puffball may produce a cloud containing as many as ______________________ spores.

19. Producing a large number of spores increases a species' chances of ______________________.

20. Fungal spores can be dispersed by ______________________, ______________________, and ______________________.

In your textbook, read about zygomycotes.

Order the steps of growth and reproduction in zygomycotes from 1 to 5.

_______ **1.** Hyphae called rhizoids penetrate the food, anchor the mycelium, and absorb nutrients.

_______ **2.** An asexual spore germinates on a food source and hyphae begin to grow.

_______ **3.** Spores are released and another asexual cycle begins.

_______ **4.** Hyphae called stolons grow across the surface of the food source and form a mycelium.

_______ **5.** Special hyphae grow upward to form sporangia that are filled with asexual spores.

Use each of the terms below just once to complete the passage.

yeasts	**conidia**	**multicellular**	**conidiophores**	**yeast cells**
sac fungi	**ascospores**	**unicellular**	**vaccine**	**ascus**

Ascomycotes are also called **(6)** _______________ because they produce sexual spores, called **(7)** _______________, in a saclike structure, called a(n) **(8)** _______________. During asexual reproduction, ascomycotes produce spores called **(9)** _______________. These asexual spores are produced in chains or clusters at the tips of structures called **(10)** _______________, which are elongated hyphae.

Morels and truffles are **(11)** _______________ ascomycotes that are edible. Yeasts are **(12)** _______________ ascomycotes. **(13)** _______________ are used to make beer, wine, and bread. They are also used in genetic research. A(n) **(14)** _______________ for the disease hepatitis B is produced from rapidly growing **(15)** _______________, which contain spliced human genes.

In your textbook, read about basidiomycotes.

Answer the following questions about the life of a mushroom.

16. What are basidia and where are they found?

____

____

____

17. What happens when mycelia of two different mating strains meet?

____

18. What does a mycelium with two nuclei in its cells form?

____

19. What does a diploid cell inside a basidium produce as a result of meiosis?

____

____

In your textbook, read about deuteromycotes, the mutualistic relationships of mycorrhizae and lichens, and the origins of fungi.

Write the letter of the item in Column B that best completes each statement in Column A.

	Column A	Column B
______	**20.** ______ is an example of a deuteromycote.	**a.** basidiomycotes
______	**21.** A mycorrhiza is a mutualistic relationship between a fungus and a(n) ______ .	**b.** mycorrhizae
______	**22.** ______ is an antibiotic produced by a deuteromycote.	**c.** alga
______	**23.** Plants that have ______ associated with their roots grow larger.	**d.** *Penicillium*
______	**24.** ______ make up a division of fungi that have no known sexual stage.	**e.** pioneer species
______	**25.** A lichen is a mutualistic relationship between a fungus and a(n) ______ or cyanobacterium.	**f.** deuteromycotes
______	**26.** Lichens are ______ in all parts of the world.	**g.** penicillin
______	**27.** Scientists think that ascomycotes and ______ evolved from a common ancestor.	**h.** plant

Name Date Class

BioDigest 6 Viruses, Bacteria, Protists, and Fungi

Reinforcement and Study Guide

In your textbook, read about viruses.

Label the parts of a virus.

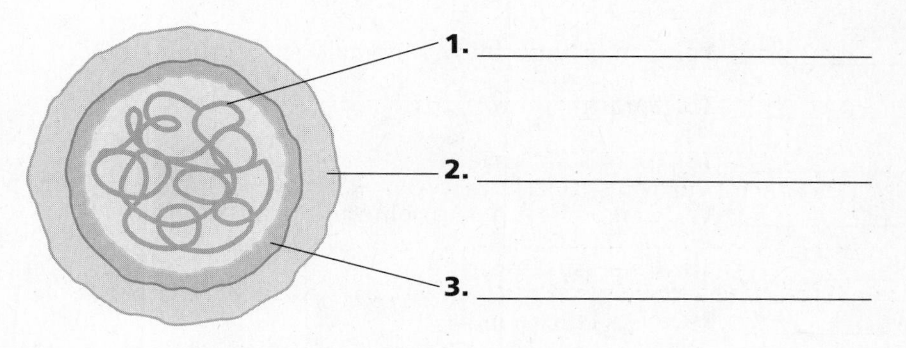

Number the following steps of the lytic cycle in the order in which they occur.

______ **4.** The viral nucleic acid causes the host cell to produce new virus particles.

______ **5.** A virus attaches to the membrane of a host cell.

______ **6.** The new virus particles are released from the host cell, killing the cell.

______ **7.** The viral nucleic acid enters the host cell.

In your textbook, read about bacteria.

Determine if the statement is true. If it is not, rewrite the italicized part to make it true.

______ **8.** A bacterium is a unicellular *eukaryote*.

______ **9.** Bacteria may be heterotrophs, photosynthetic autotrophs, or *chemosynthetic autotrophs*.

______ **10.** Bacteria reproduce asexually by *conjugation*.

______ **11.** Bacteria that are *obligate aerobes* are killed by oxygen.

______ **12.** *Archaebacteria* often live in extreme environments.

______ **13.** Some bacteria fix *oxygen*.

BioDigest 6

Viruses, Bacteria, Protists, and Fungi, *continued*

Reinforcement and Study Guide

In your textbook, read about protists.

Write the letter of the item in Column B that best matches the item in Column A.

	Column A	Column B
______	**14.** Can be unicellular, colonial, or multicellular	**a.** green algae
______	**15.** Parasitic protozoans	**b.** sporozoans
______	**16.** Causes malaria	**c.** amoeba
______	**17.** Can be both autotrophic and heterotrophic	**d.** slime mold
______	**18.** Funguslike protist	**e.** brown algae
______	**19.** Uses cilia to move	**f.** *Plasmodium*
______	**20.** Contain carotenoids	**g.** euglenas
______	**21.** Have hard, armorlike plates	**h.** *Paramecium*
______	**22.** Kelps	**i.** dinoflagellates
______	**23.** Uses pseudopodia to move	**j.** diatoms
______	**24.** Have red and blue pigments	**k.** red algae

In your textbook, read about fungi.

Answer the following questions.

25. How do fungi obtain nutrients from a food source? ______________________________

____

26. How do fungi play a role in recycling nutrients on Earth? ______________________________

____

27. What are hyphae? ______________________________

28. In what structure do club fungi produce sexual spores? ______________________________

29. In what structure do sac fungi produce sexual spores? ______________________________

30. How do mycorrhizae benefit both the plants and the fungi? ______________________________

____

____

____

31. What is a lichen? ______________________________

____

Name _______________ Date _______ Class _______

Chapter 21

What Is a Plant?

Reinforcement and Study Guide

Section 21.1 Adapting to Life on Land

In your textbook, read about the origins and adaptations of plants.

For each answer given below, write an appropriate question.

1. **Answer:** Multicellular eukaryotes having thick cell walls made of cellulose, a protective, waterproof covering, and that can produce its own food in the form of glucose through photosynthesis

   Question: _______________

2. **Answer:** The earliest known plant fossils

   Question: _______________

3. **Answer:** Protective, waxy layers that cover most fruits, leaves, and stems

   Question: _______________

4. **Answer:** The organ of a plant that traps light energy for photosynthesis, and is supported by a stem

   Question: _______________

5. **Answer:** The organ that works like an anchor, transports nutrients, and absorbs water and minerals

   Question: _______________

In your textbook, read about alternation of generations.

Use each of the terms below just once to complete the passage.

diploid	**generations**	**meiosis**
gametes	**haploid**	**sporophyte**

The lives of all plants consist of two alternating stages, or **(6)** _______________ . The gametophyte generation of a plant is responsible for the development of **(7)** _______________ . All seeds of the gametophyte, including the gametes, are **(8)** _______________ . The **(9)** _______________ generation is responsible for the production of spores. All cells of the sporophyte are **(10)** _______________ . The spores are produced by the sporophyte plant body by **(11)** _______________ and are, therefore, haploid.

In your textbook, read about the origin and adaptations of plants.

Circle the letter of the choice that best completes the statement.

12. The lives of ________ plants include two generations that alternate.

a. non-seed producing **b.** seed
c. all **d.** most

13. The generation of a plant responsible for producing gametes is the

a. alternation of generations. **b.** gametophyte generation.
c. sporophyte generation. **d.** seed-producing generation.

14. All gametophyte spores are ________ and all sporophyte tissue cells are ________ .

a. haploid/diploid. **b.** diploid/haploid.
c. haploid/haploid. **d.** diploid/diploid.

15. Non-seed plants ________ that grow into gametophytes.

a. release spores into the environment **b.** retain spores in the parent plant
c. release seeds into the environment **d.** retain seeds in the parent plant

Answer the following questions.

16. What is the difference between vascular and nonvascular plants?

17. Some land plants produce seeds. What is their function? How do they differ from spores?

18. How do algae and land plants get nutrients?

Section 21.2 Survey of the Plant Kingdom

In your textbook, read about non-seed plants.

For each item in Column A, write the letter of the matching item in Column B.

	Column A	Column B
______	**1.** Leaves that are found on ferns	**a.** leafy liverworts
______	**2.** Scaly structures that support male or female reproductive structures	**b.** thallose liverworts
______	**3.** Plants with a broad, flattened body that resembles a lobed leaf	**c.** fronds
______	**4.** Plants with three flattened rows of thin leaves	**d.** hornworts
______	**5.** Nonvascular plants that grow in damp, shady habitats and whose sporophytes resemble horns	**e.** cones

Complete the chart below by marking the appropriate columns for each division of plants.

Division	Vascular	Nonvascular	Non-seed Plants	Seeds in Fruits	Seeds in Cones
6. Hepatophyta					
7. Anthocerophyta					
8. Bryophyta					
9. Psilophyta					
10. Lycophyta					
11. Sphenophyta					
12. Pterophyta					
13. Cycadophyta					
14. Gnetophyta					
15. Ginkgophyta					
16. Coniferophyta					
17. Anthophyta					

Section 21.2 Survey of the Plant Kingdom, *continued*

In your textbook, read about non-seed plants.

For each answer given below, write an appropriate question.

18. Answer: Vascular plants that have neither roots nor leaves

Question: ______________________________

19. Answer: Vascular plants that have hollow, jointed stems surrounded by whorls of scalelike leaves, whose cells contain large deposits of silica

Question: ______________________________

20. Answer: Plants that may be the ancestors of all plants

Question: ______________________________

21. Answer: Hard-walled reproductive cells found in non-seed plants

Question: ______________________________

22. Answer: Nonvascular plants that rely on osmosis and diffusion to transport water and nutrients, although some members have elongated cells that conduct water and sugars

Question: ______________________________

In your textbook, read about seed plants.

Use each of the terms below just once to complete the passage.

Anthophyta	**Cycadophyta**	**Gnetophyta**
Coniferophyta	**Ginkgophyta**	

Seed plants are classified into five divisions. Plants from the **(23)** ________________ division are palmlike trees with scaly trunks and are often mistaken for ferns or small palm trees. There is only one living species in the **(24)** ________________ division. The members of the **(25)** ________________ division are the largest, most diverse group of seed plant on Earth and are commonly known as the flowering plants. Three distinct genera make up the plant division called **(26)** ________________. Species of the **(27)** ________________ division can be identified by the characteristics of their needlelike or scaly leaves.

Name Date Class

Chapter 22 **The Diversity of Plants**

Reinforcement and Study Guide

Section 22.1 Nonvascular Plants

In your textbook, read about nonvascular plants—bryophyta, hepatophyta, and anthocerophyta.

Complete each statement.

1. Nonvascular plants are successful in habitats with adequate ________________ .

2. The ________________ generation is dominant in nonvascular plants.

3. Sperm are produced in male reproductive structures called ________________ , and eggs are produced in female reproductive structures called ________________ .

4. Mosses have colorless multicellular structures called ________________ , which help anchor the stem to the soil.

5. Most liverworts have ________________________ that helps reduce evaporation of water from the plant's tissues.

6. Liverworts occur in many environments and include two groups: the ________________ liverworts and the ________________ liverworts.

7. One unique feature of hornworts is the presence of a(n) ______________________ in each cell.

8. The common names for the nonvascular plants, Bryophyta, Hepatophyta, and Anthocerophyta are ________________ , ________________ , and ________________ .

Circle the letter of the response that best completes the statement.

9. Nonvascular plants are not as common or as widespread as vascular plants because

a. nonvascular plants are small.

b. the life functions of nonvascular plants require a close association with water.

c. nonvascular plants are limited to dry habitats.

d. none of the above.

10. The life cycle of nonvascular plants includes an alternation of generations between a

a. diploid sporophyte and a diploid gametophyte.

b. haploid sporophyte and a haploid gametophyte.

c. diploid sporophyte and a haploid gametophyte.

d. haploid sporophyte and a diploid gametophyte.

11. Fossil and genetic evidence suggests that the first land plants were

a. mosses.

b. sphagnum moss.

c. liverworts.

d. hornworts.

Chapter 22 **The Diversity of Plants,** *continued*

Reinforcement and Study Guide

Section 22.2 Non-Seed Vascular Plants

In your textbook, read about the alternation of generations of non-seed vascular plants and lycophyta.

Use each of the terms below just once to complete the passage.

antheridia	**leaves**	**sporophyte**
archegonia	**prothallus**	**strobilus**
egg	**reproductive cells**	**zygote**
fertilization	**sperm**	

Unlike nonvascular plants, the spore-producing **(1)** _______________ is the dominant generation in vascular plants. A major advance in vascular plants was the adaptation of **(2)** _______________ to form structures that protect the developing **(3)** _______________ . In some non-seed vascular plants, spore-bearing leaves form a compact cluster called a(n) **(4)** _______________. Spores are released from this compact cluster. These spores then grow to form the gametophyte, called a(n) **(5)** _______________ . This structure is relatively small and lives in or on soil. The prothallus then forms **(6)** _______________ , male reproductive structures, and **(7)** _______________ , female reproductive structures. **(8)** _______________ are released from the antheridium and swim through a film of water to the **(9)** _______________ in the archegonium. **(10)** _______________ occurs and a large, dominant sporophyte plant develops from the fertilized **(11)** _______________ .

For each statement below, write true or false.

__________ **12.** The leafy stems of lycophytes resemble clubs, and their reproductive structures are moss shaped.

__________ **13.** The leaves of lycophytes occur as pairs, whorls, or spirals along the stem.

__________ **14.** Lycophytes are simple vascular plants with creeping leaves.

__________ **15.** The club moss is commonly called ground pine because it is evergreen and resembles a miniature pine tree.

Chapter 22 The Diversity of Plants, *continued*

Reinforcement and Study Guide

Section 22.2 Non-Seed Vascular Plants, continued

In your textbook, read about sphenophyta and pterophyta.

Complete each statement.

16. The hollow-stemmed horsetail appears to be jointed with scalelike ________________ surrounding each joint.

17. The most recognized generation of ferns is the ________________ generation.

18. The ________________ in most ferns is a thin, flat structure.

19. In most ferns, the main stem, called a ________________ , is underground. It contains many starch-filled cells for ________________ .

20. The leaves of a fern are called ________________ and grow upward from the rhizome.

21. Fronds are often divided into leaflets called ________________ , which are attached to a central stipe.

22. Ferns were the first vascular plants to evolve leaves with branching ________________ of vascular tissue.

23. The common names for the seedless vascular plants, Lycophyta, Sphenophyta, and Pterophyta are ________________ , ________________ , and ________________ .

Answer the following questions on the lines provided.

24. Why are sphenophytes, or horsetails, sometimes referred to as scouring rushes?

____

____

____

25. Why might you be more familiar with ferns than with club mosses and horsetails?

____

____

____

____

Section 22.3 Seed Plants

In your textbook, read about the seed plants—cycadophyta, gingkophyta, gnetophyta, coniferophyta, and anthophyta.

Complete each statement.

1. An ________________, or young diploid sporophyte, has food-storage organs called ________________, which develop into leaves.

2. Vascular plants that produce ________________ in cones are sometimes called ________________.

3. Seed plants do not require ________________ for ________________.

4. The male gametophyte develops inside a structure called a(n) ________________ that includes sperm cells, nutrients, and a protective outer covering.

5. The female gametophyte, which produces the egg cell, is contained within a sporophyte structure called a(n) ________________.

6. Biennials develop large storage roots and live for ________________.

7. Perennials produce flowers and seeds periodically for ________________.

8. Annual plants live for ________________.

9. ________________ have one seed leaf; ________________ have two seed leaves.

For each statement below, write true or false.

__________ **10.** Cycads are related to palm trees but their leaves unfurl like fern fronds.

__________ **11.** There is only one species of ginkgo tree alive today.

__________ **12.** Most gnetophytes today are found in the deserts or mountains of Africa, Asia, North America, and Central or South America.

__________ **13.** Most conifers are evergreen plants that lose their needlelike leaves all at once and only grow in nutrient-rich soil.

__________ **14.** Dropping leaves is an adaptation in deciduous plants to reduce water loss when it is less available during winter.

__________ **15.** Anthophytes are unique in that they are the only division of plants that produce fruits.

Name _______________ Date _______ Class _______

Chapter 23

Plant Structure and Function

Reinforcement and Study Guide

Section 23.1 Plant Cells and Tissues

In your textbook, read about plant cells and tissues.

Match the definitions in Column 1 with the terms in describes from Column 2. Place the letter from Column 2 in the spaces under Column 1.

Column 1

_______ **1.** The most abundant kind of plant cells

_______ **2.** Long cells with unevenly thickened cell walls. This type of cell wall allows the cells to grow.

_______ **3.** Cells with walls that are very thick and rigid. At maturity, these cells often die, leaving the cell walls to provide support for the plant.

_______ **4.** Dermal tissue that is composed of flattened parenchyma cells that cover all parts of the plant

_______ **5.** Openings in the cuticle of the leaf that control the exchange of gases

_______ **6.** Cells that control the opening and closing of the stomata.

_______ **7.** Hairlike projections that extend from the epidermis

_______ **8.** Plant tissue composed of tubular cells that transports water and minerals from the roots to the rest of the plant

_______ **9.** Tubular cells, with tapered ends, which transport water throughout a plant

_______ **10.** Lateral meristem that produces a tough covering for the surface of stems and roots

_______ **11.** Vascular tissue that transport sugars from the leaves to all parts of the plant

_______ **12.** Long, cylindrical phloem cells through which sugars and organic compounds flow

_______ **13.** Nucleated cells that help manage the transport of sugars and other organic compounds through the sieve cells of the phloem

_______ **14.** Areas where new cells are produced

_______ **15.** Growth tissue found at or near the tips of roots and stems

_______ **16.** Tubular cells that transport water throughout the plant. These cells are wider and shorter than tracheids.

_______ **17.** Lateral meristem that produces new xylem and phloem cells in the stems and roots

Column 2

a. apical meristem

b. collenchyma

c. companion cell

d. cork cambium

e. epidermis

f. guard cells

g. meristem

h. parenchyma

i. phloem

j. sclerenchyma

k. sieve tube member

l. stomata

m. tracheids

n. trichomes

o. vascular cambium

p. vessel element

q. xylem

Chapter 23 **Plant Structure and Function,** *continued*

Reinforcement and Study Guide

Section 23.2 Roots, Stems, and Leaves

In your textbook, read about roots and stems.

Label the parts of the dicot root. Use these choices:

cortex **phloem** **epidermis** **endodermis** **xylem**

Dicot Root

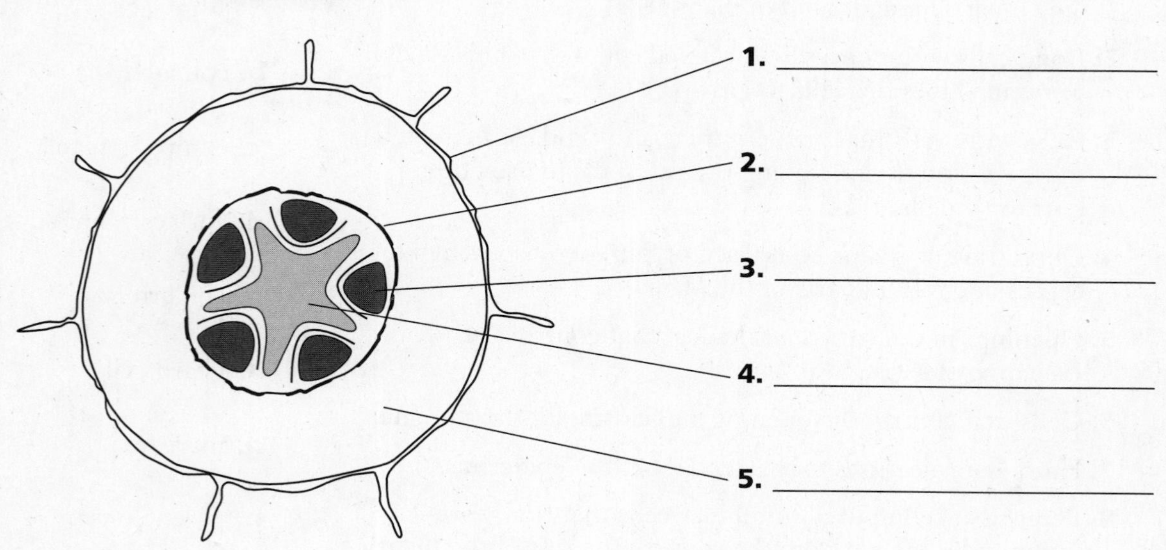

For each statement below, write true or false.

______ **6.** A root hair is a small extension of an epidermal, or outermost, cell layer of a dicot root.

______ **7.** Layers of parenchyma cells make up the cortex of a dicot root and the central pith of a monocot root.

______ **8.** Outside the endodermis is a tissue called the pericycle that develops vertical roots.

______ **9.** Vascular cambium cells found at the center of a root grow more xylem and phloem cells that increase the size of the root.

______ **10.** Behind the root tip are cell-producing growth tissues called the root cap.

______ **11.** The difference between roots and stems lies in the way they transport water.

______ **12.** Primary growth in a stem occurs in the apical meristem.

Plant Structure and Function, *continued*

Section 23.2 Roots, Stems, and Leaves, continued

In your textbook, read about stems and leaves.

Circle the letter of the response that best completes the statement.

13. Many wildflowers with soft, green stems are plants that have
- **a.** woody stems.
- **b.** herbaceous stems.
- **c.** woody roots.
- **d.** all of the above.

14. The functions of a plant's stem include
- **a.** transporting sugar.
- **b.** supporting the plant.
- **c.** transporting water and minerals.
- **d.** all of the above.

15. Any portion of the plant that stores sugars is called a
- **a.** petiole.
- **b.** mesophyll.
- **c.** root cap.
- **d.** sink.

16. The movement of sugars from the leaves through the phloem is called
- **a.** photosynthesis.
- **b.** transpiration.
- **c.** translocation.
- **d.** food storage.

In your textbook, read about the leaves of a plant.

Use each of the terms below just once to complete the passage.

stomata	**extend**	**cuticle**	**transpiration**	**epidermis**
veins	**stem**	**petiole**	**photosynthesis**	**mesophyll**

There are many parts to a leaf. Grass leaves grow right out of the **(17)** ______________________, but other leaves are connected to the stem by a stalk called the **(18)** ______________________. The petiole is made of vascular tissues that **(19)** ______________________ up into the leaf to form **(20)** ______________________.

The outer surface of a leaf has a **(21)** ______________________ that covers the epidermis. Inside the epidermis are two layers of photosynthetic cells that make up the **(22)** ______________________. Cells in the palisade layer have many chloroplasts and carry out most of the leaf's **(23)** ______________________. Leaves have a(n) **(24)** ______________________ with a waxy cuticle and **(25)** ______________________ help prevent water loss. The loss of water through the stomata is called **(26)** ______________________.

Chapter 23 **Plant Structure and Function,** *continued*

Reinforcement and Study Guide

Section 23.3 Plant Responses

In your textbook, read about plant hormones and plant responses.

Complete each statement.

1. A ______________________________ is a chemical that is produced in one part of an organism and transported to another part, where it causes a physiological change.

2. The group of plant hormones called ______________________________ promote cell elongation. Indoleacetic acid (IAA) is an example of this group of hormones.

3. The group of growth hormones that cause plants to grow taller because, like auxins, they stimulate cell elongation, are called ______________________________ .

4. The hormones called ______________________________ are so named because they stimulate cell division by stimulating the production of proteins needed for mitosis.

5. The plant hormone called ______________________________ is a simple, gaseous compound composed of carbon and hydrogen that speeds the ripening of fruits.

6. A plant's response to an external stimulus that comes from a particular direction is called a ______________________________ .

7. A responsive movement of a plant that is not dependent on the direction of the stimulus is called a ______________________________ .

Determine if the statement is true. If it is not, rewrite the italicized part to make it true.

8. A *large* amount of hormone is needed to make physiological changes in a plant.

____

9. If gibberellins are applied to the tip of a dwarf plant, it will grow *taller*.

____

10. The growth of a plant towards light is caused by an unequal distribution of *ethylene* in the plant's stem.

____

11. If a tropism is *negative*, the plant grows toward the stimulus.

____

12. The growth of a plant toward light is called *phototropism*.

____

13. *Gravitropism* is the direction of plant growth in response to gravity.

____

14. A plant's response to touch is called *cytokinin*.

____

Name Date Class

Chapter 24

Reproduction in Plants

Reinforcement and Study Guide

Section 24.1 Life Cycles of Mosses, Ferns, and Conifers

In your textbook, read about alternation of generations and the life cycles of mosses and ferns.

Use each of the terms below just once to complete the following statements.

diploid	**meiosis**	**sporophyte**
dominant	**mitosis**	**vegetative reproduction**
egg	**protonema**	
gametophyte	**sperm**	

1. The two phases of the plant life are the ______________________________ stage and the ______________________________ stage.

2. The cells of the sporophyte are all ______________________________ .

3. The female gamete is the ____________________ , and the male gamete is the ____________________ .

4. Some plants reproduce asexually by a process called ______________________________ , in which a new plant is produced from an existing vegetative structure.

5. Mosses belong to one of the few plant divisions in which the gametophyte plant is the ______________________________ generation.

6. A small, green filament of moss cells that develops into either a male or female moss gametophyte is known as a(n) ______________________________ .

7. The moss diploid zygote divides by ______________________________ to form a new sporophyte in the form of a stalk and capsule.

8. Spores are produced by ______________________________ in the capsule of the moss sporophyte.

Number each description to order the stages from spore release of the life cycle of a fern, from 1 to 7.

____________ **9.** A spore germinates and grows into a heart-shaped gametophyte called a prothallus.

____________ **10.** After fertilization, the diploid zygote grows into a sporophyte.

____________ **11.** As the sporophyte grows, roots and fronds grow out from the rhizome.

____________ **12.** Sperm swim through a film of water on the prothallus to reach and fertilize an egg in the archegonium.

____________ **13.** In each sporangium, spores are produced by meiosis, and the cycle begins again as the spores are dispersed by the wind.

____________ **14.** Sori, or clusters of sporangia, grow on the pinnae.

____________ **15.** A sporangium bursts, releasing haploid spores.

Chapter 24 **Reproduction in Plants,** *continued*

Reinforcement and Study Guide

Section 24.1 Life Cycles of Mosses, Ferns, and Conifers, continued

In your textbook, read about the life cycle of conifers.

Answer the following questions.

1. What is the dominant stage in conifers?

2. What does the adult conifer produce on its branches?

3. What is a megaspore?

4. What are microspores and how are they produced?

5. What do the microspores develop into?

6. What is a micropyle?

7. How does fertilization take place?

8. After fertilization, a zygote develops inside the ovule into an embryo with several cotyledons. What happens to the ovule?

9. What happens to the seeds when the female cone opens and falls to the ground?

10. What will the seedling become?

Section 24.2 Flowers and Flowering

In your textbook, read about the structure of a flower.

Determine if the statement is true. If it is not, rewrite the italicized part to make it true.

1. In flowering plants, sexual reproduction takes place in the *seed*, which has several parts.

____

2. The structure of a flower includes four kinds of organs: *sepals, petals, stamens, and ovaries.*

____

3. *Petals* are usually colorful, leaflike structures that encircle the flower stem.

____

4. The male reproductive structure located inside the petals of a flower is a *stamen*. Sperm-containing pollen is produced in the *anther* at the tip of the stamen.

____

5. The female reproductive structure at the center of a flower is the *ovary*. Eggs are formed in the *pistil*, which is located in the bottom portion of the *ovary*.

____

Label the parts of the flower. Use these choices:

sepal	**petal**	**stigma**	**ovule**
anther	**ovary**	**filament**	

6. ______________

7. ______________

8. ______________

9. ______________

10. ______________

11. ______________

12. ______________

Section 24.3 The Life Cycle of a Flowering Plant

In your textbook, read about the life cycle of a flowering plant.

For each item in Column A, write the letter of the matching item from Column B.

	Column A	Column B
______	**1.** Two nuclei in one cell at the center of the embryo sac	**a.** dormancy
______	**2.** A process in which one sperm fertilizes the egg and the other sperm joins with the central cell	**b.** double fertilization
______	**3.** Food-storing tissue that develops from the triploid central cell and supports the development of the embryo	**c.** endosperm
______	**4.** A period of inactivity in which seeds of some plant species remain until conditions are favorable for growth and development	**d.** germination
______	**5.** The beginning of the development of the embryo into a new plant	**e.** hypocotyl
______	**6.** This embryonic root is the first part of the embryo to appear from the seed	**f.** polar nuclei
______	**7.** The portion of the stem near the seed	**g.** radicle

Answer the following questions.

8. How do anthophytes attract animal pollinators?

9. How do seeds form after fertilization takes place?

10. Name three ways seeds are dispersed.

BioDigest 7 **Plants**

Reinforcement and Study Guide

In your textbook, read about plants.

Study the following diagram of alternation of generations in plants. Then answer the questions.

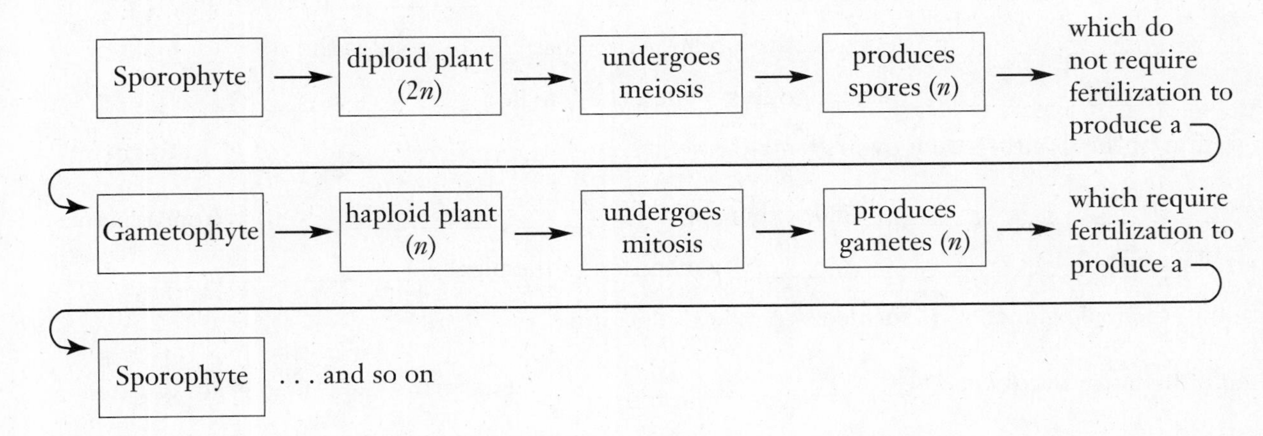

1. Is the sporophyte generation a haploid or diploid generation? ______

2. Is the gametophyte generation a haploid or diploid generation? ______

3. Is a spore haploid or diploid? ______

4. Is a gamete haploid or diploid? ______

5. Does a spore produce a gametophyte without fertilization? ______

6. Does a gamete produce a sporophyte without fertilization? ______

7. Which generation—sporophyte or gametophyte—produces a generation that is diploid?

______

Seed Plants

Explain how these adaptations enable conifers to survive in cold or dry climates.

8. *Needles* ______

______

9. *Stems* ______

______

10. *Flexible leaves and branches* ______

______

______

Flowering Plants

Fill in the following blanks to explain the function of a flower.

11. A flower has two major reproductive structures. The ______________________ is the ______________________ reproductive organ. At the base of the pistil is the ______________________, which houses ovules, the female ______________________ generation of the plant. In each ovule, female gametes, or ______________________, form.

12. The ______________________ and anther form the ______________________ reproductive organ. The male ______________________ generation of the plant is ______________________. Within it, the male gametes are formed.

13. When pollination occurs, a ______________________ extends from the pollen grain to the ovary, and two ______________________ travel down the tube to fertilize the eggs in the ovule.

14. Some flowers are colorful and showy. Others are small and inconspicuous. Explain how these two flower types are adapted to different pollinators.

____

____

____

15. Discuss three ways that seeds may be spread through the environment.

a. ____

b. ____

c. ____

Name ______ Date ______ Class ______

Chapter 25

What Is an Animal?

Reinforcement and Study Guide

Section 25.1 Typical Animal Characteristics

In your textbook, read about the characteristics of animals.

Answer the following questions.

1. You have just discovered a new organism that you think is an animal. In order to be classified as an animal, what characteristics must it have?

____

____

2. What is one important factor that influences how an animal obtains its food?

____

3. How might an animal be free-moving at one stage in its life and sessile at another? Give an example.

____

____

4. How do sessile, aquatic animals get their food?

____

____

If the animal described below is a sessile organism, write yes. If it is not, write no.

______________ **5.** Barnacles attached to a ship's hull

______________ **6.** A spider lying in wait in the center of its web

______________ **7.** Coral larvae drifting in a tropical ocean

______________ **8.** Sponges growing on the outside of a crab's shell

Complete each statement.

9. Digestion in a sponge takes place in ______________ ______________, while digestion in a more complex animal like a tiger takes place in a(n) ______________ ______________.

10. Some of the food you had for breakfast has been stored as ______________ or ______________, ready to be used if it's a long time until your next meal.

Chapter 25 **What Is an Animal?, *continued***

Reinforcement and Study Guide

Section 25.1 Typical Animal Characteristics, continued

In your textbook, read about the development of animals.

Determine if the statement is true. If it is not, rewrite the italicized part to make it true.

11. Most animals develop from a single, fertilized egg called a *blastula*. ______________________

12. A zygote divides by a process known as *cleavage*. ______________________

13. The blastula is a *solid ball of cells*. ______________________

____

Label the parts of the gastrula shown here. Use these choices:

ectoderm endoderm mesoderm opening in gastrula

14. ______________________

15. ______________________

16. ______________________

17. ______________________

Complete the chart by checking the correct column for each description.

Description	Endoderm	Ectoderm	Mesoderm
18. Gives rise to digestive tract			
19. Continues to grow and divide			
20. Lines the inner surface of gastrula			
21. Gives rise to muscles			
22. Develops into skin and nervous tissue			
23. Forms from cells that break off endoderm			

Section 25.2 Body Plans and Adaptations

In your textbook, read about kinds of symmetry in animals.

Circle the letter of the choice that best completes the statement or answers the question.

1. Different kinds of symmetry make it possible for animals to
- **a.** grow very large.
- **b.** survive when cut into pieces.
- **c.** move and find food in different ways.
- **d.** live a long time.

2. The irregularly shaped body of a sponge is an example of
- **a.** asymmetry.
- **b.** gastrulation.
- **c.** symmetry.
- **d.** balance.

3. A sponge's body has how many layers of cells?
- **a.** one
- **b.** two
- **c.** three
- **d.** four

4. The embryonic development of a sponge does *not* include which of the following?
- **a.** formation of endoderm
- **b.** formation of mesoderm
- **c.** a gastrula stage
- **d.** a, b, and c

5. If you divided a radially symmetrical animal along any plane through its central axis, you would end up with
- **a.** roughly equal halves.
- **b.** front and back halves.
- **c.** top and bottom halves.
- **d.** three pieces.

6. Which of the following animals is *not* radially symmetrical?
- **a.** a hydra
- **b.** a sea urchin
- **c.** a spider
- **d.** a starfish

7. An organism with bilateral symmetry can be divided lengthwise into right and left halves that are
- **a.** asymmetrical.
- **b.** mirror images of each other.
- **c.** made up of two cell layers.
- **d.** flattened.

Identify each of the following body parts as being either dorsal or ventral on the animal's body.

_______________ **8.** the navel of a killer whale

_______________ **9.** the sail fin on an iguana

_______________ **10.** the back of your neck

_______________ **11.** the mouth of a shark

_______________ **12.** the pouch of a kangaroo

In your textbook, read about bilateral symmetry and body plans.

Answer the following questions.

13. In what ways was the development of a body cavity, or coelom, an advantage for bilaterally symmetrical animals?

____

____

14. Describe an acoelomate animal's body plan.

____

____

15. How do nutrients get to the cells in a flatworm's solid, acoelomate body?

____

Use each of the terms below just once to complete the passage.

coelom	**completely**	**double**	**internal organs**
mesoderm	**partly**	**pseudocoelom**	

A roundworm has a **(16)** ________________, a fluid-filled body cavity that is **(17)** ________________ lined with **(18)** ________________. Coelomate animals have a **(19)** ________________, a body cavity that is **(20)** ________________ surrounded by mesoderm and in which complex **(21)** ________________ are suspended by **(22)** ________________ layers of mesoderm tissue.

In your textbook, read about animal protection and support.

For each statement below, write <u>true</u> or <u>false</u>.

________________ **23.** During the course of evolution, animal body plans have decreased in complexity.

________________ **24.** An exoskeleton provides protection and support on the outside of an animal's body, as well as a place for muscle attachment.

________________ **25.** An endoskeleton is a support framework housed within the body, a protective enclosure for internal organs, and a brace for muscles to pull against.

________________ **26.** An invertebrate is an animal with a backbone.

Name Date Class

Chapter 26

Sponges, Cnidarians, Flatworms, and Roundworms

Reinforcement and Study Guide

Section 26.1 Sponges

In your textbook, read about sponges.

Answer the following questions.

1. How does the name *Porifera* relate to the structure of a sponge?

____

____

2. How do sponges obtain food from their environment?

____

____

3. Describe a sponge's body plan.

____

____

Complete the table by writing a cell type or structure in sponges that fits each description.

Type of Cell or Structure	Description
4.	Aid in reproduction and nutrient transport Help produce spicules
5.	Form the outside surface of body Contract to close pores
6.	Line interior of sponge's body Use flagella to draw water through pores
7.	Found in jellylike substance between layers Make up sponge's support system

Use each of the terms below just once to complete the passage.

external buds	**eggs**	**hermaphroditic**	**internal fertilization**
larvae	**sexual**	**sperm**	

Sponges sometimes reproduce asexually by forming **(8)** ________________ . Being **(9)**________________ , a sponge can also produce both **(10)** ________________ and sperm. During **(11)** ________________ reproduction, **(12)** ________________ from one sponge fertilize the eggs of another. Fertilization can be external, but **(13)** ________________ is more common. Free-swimming **(14)** ________________ settle and develop into sessile adults.

Chapter 26

Sponges, Cnidarians, Flatworms, and Roundworms, *continued*

Reinforcement and Study Guide

Section 26.2 Cnidarians

In your textbook, read about cnidarians.

Identify each of the following descriptions as either the polyp or medusa form of a cnidarian.

______ **1.** Reef-building corals on the Great Barrier Reef

______ **2.** *Aurelia*, the moon jellyfish

______ **3.** Deep sea anemones with meter-long tentacles

______ **4.** The asexual phase in a jellyfish's life cycle

Answer the following questions.

5. Nematocysts are characteristic of cnidarians. How does a nematocyst work?

6. Compare and contrast how food is digested in a sponge and in a cnidarian.

7. How does a nerve net function?

Order the following steps in the life cycle of a jellyfish from A to F, beginning with the release of eggs and sperm.

______ **8.** A polyp grows and buds repeatedly.

______ **9.** External fertilization takes place in the sea.

______ **10.** A zygote develops into a blastula, which develops into a larva.

______ **11.** Male and female medusae release sperm and eggs, respectively.

______ **12.** A cilia-covered larva settles onto a surface.

______ **13.** A tiny medusa breaks free from its sessile parent and drifts away.

Name Date Class

Chapter 26

Sponges, Cnidarians, Flatworms, and Roundworms, *continued*

Reinforcement and Study Guide

Section 26.3 Flatworms

In your textbook, read about flatworms.

For each statement below, write <u>true</u> or <u>false</u>.

________ **1.** Flatworms are bilaterally symmetrical and have a clearly defined head.

________ **2.** Adult planarians can focus well enough with their eyespots to form images of objects in their environment.

________ **3.** Flame cells play an important role in maintaining water balance in planaria.

________ **4.** A planarian uses its pharynx to locate food.

________ **5.** Planarians reproduce sexually by producing encapsulated zygotes that hatch into free-swimming larvae.

In part C of the illustration below, draw in what you think will happen to the two halves of the cut planarian. Then, answer the question.

6.

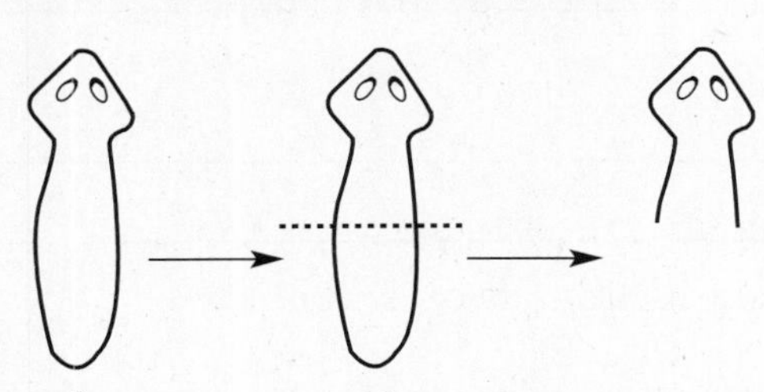

7. How is regeneration adaptive for survival in planarians?

____

____

____

____

Complete the table by checking the correct column for each description.

Description	Planarian	Tapeworm	Fluke
8. Lives parasitically within a host			
9. Body made up of proglottids			
10. Body is thin and solid			
11. Free-living in aquatic environments			
12. Attaches to host's intestine with scolex			
13. Extends a pharnyx to suck up food			
14. May live in host's blood vessels			

Chapter 26

Sponges, Cnidarians, Flatworms, and Roundworms, *continued*

Reinforcement and Study Guide

Section 26.4 Roundworms

In your textbook, read about roundworms.

Answer the following questions.

1. What impact do parasitic roundworms have on other organisms?

____

____

2. List three ways in which roundworms differ from flatworms.

____

____

3. What accounts for the characteristic wriggling movement of roundworms?

____

____

4. What are four of the most common parasitic roundworms that infect humans?

____

____

5. Can roundworms cause plant diseases? Explain.

____

____

6. What parts of plants are most commonly susceptible to parasitic roundworms?

____

____

Below are two medical reports. After reading each report, give a preliminary diagnosis of what you think might be causing the problem.

7.

MEDICAL REPORT

Patient is an active 5-year-old girl. Complains about a constant itching around the anal area, especially at night.

Preliminary Diagnosis:

____

8.

MEDICAL REPORT

Patient is a 29-year-old female Peace Corps volunteer. Lived with remote tribe whose primary food is pigs. Complains of muscle pain.

Preliminary Diagnosis:

____

Name Date Class

Chapter 27 Mollusks and Segmented Worms

Reinforcement and Study Guide

Section 27.1 Mollusks

In your textbook, read about what a mollusk is.

The phylum *Mollusca* is a very diverse group of animals. Complete the table by checking the correct column for each characteristic.

Characteristic	Exhibited In:	
	All Mollusks	**Some Mollusks**
1. Possess a hard, external shell		
2. Bilaterally symmetrical		
3. Have a mantle		
4. Live on land		
5. Digestive tract has two openings.		
6. Inhabit aquatic environments		
7. Share similar developmental patterns		
8. Are slow-moving		
9. Have a coelom		

In your textbook, read about diversity of mollusks.

Identify each mollusk shown below. Write the name of the class to which it belongs and briefly describe where it lives.

10.

____
____
____
____
____

11.

____
____
____
____
____

12.

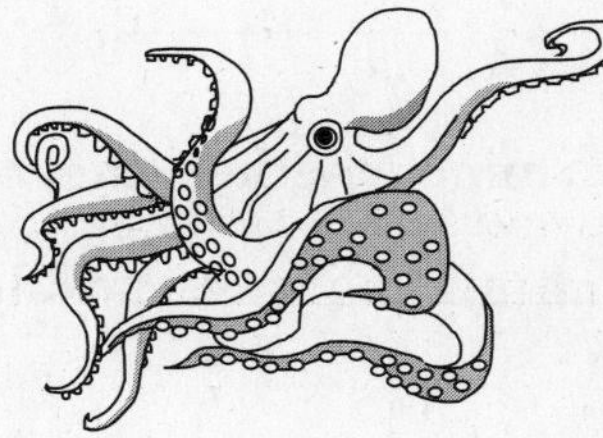

____
____
____
____
____

Chapter 27 **Mollusks and Segmented Worms,** *continued*

Reinforcement and Study Guide

Section 27.1 Mollusks, continued

In your textbook, read about a mollusk's body systems and the diversity of mollusks.

Complete each statement.

13. Gastropods have either a(n) ________________ shell or ________________ shell.

14. Most mollusks have a(n) ________________ circulatory system in which blood flows through ________________ into open ________________ around tissues and organs.

15. Most mollusks use ________________ for respiration, while a garden slug uses a primitive ________________ for gas exchange.

16. ________________ are involved in removing wastes from a mollusk's body.

17. Fertilization in most aquatic mollusks takes place ________________ .

For each statement below, write true or false.

________________ **18.** All shelled gastropods are predators.

________________ **19.** When a snail is disturbed, it pulls its body inside its shell for protection.

________________ **20.** Without a shell, terrestrial slugs and sea slugs (nudibranchs) have no protection against predators.

________________ **21.** The two shells of bivalve mollusks are held together by the mantle.

Determine whether each of the statements below best describes bivalves, gastropods, or both.

________________ **22.** Nearly all feed by filtering particles from the water around them.

________________ **23.** Most have a large muscular foot.

________________ **24.** They use a radula for feeding.

________________ **25.** Water flows through their bodies via well-developed incurrent and excurrent siphons.

In your textbook, read about cephalopods.

Answer the following questions.

26. Describe the "head-foot" region of a cephalopod.

____

____

____

____

27. What would you expect to find on the interior surfaces of a squid's many arms?

____

28. How does the intelligence of an octopus compare to that of a clam?

____

____

____

Using what you know about the three major classes of mollusks, complete the chart below by checking the correct column(s) for each characteristic.

Characteristics	Type of Mollusk		
	Gastropods	**Bivalves**	**Cephalopods**
29. Intelligent, with a well-developed nervous system			
30. Have no distinct head			
31. Have an open circulatory system			
32. External shells present in some species			
33. All species are carnivorous predators.			
34. Use a radula in feeding			
35. All use gills for both respiration and food collection.			
36. Bite prey with a beak			

Chapter 27

Mollusks and Segmented Worms, *continued*

Section 27.2 Segmented Worms

In your textbook, read about segmented worms, including the Inside Story about earthworms.

Use each of the terms below just once to complete the passage:

Annelida	**bristleworms**	**earthworms**	**muscles**
parapodia	**segments**	**setae**	

Members of the phylum **(1)** ________________ all have bodies made up of multiple **(2)** ________________ . Each segment has its own **(3)** ________________ that function to lengthen and shorten the worm's body. When present, bristlelike **(4)** ________________ act as anchors while the worm is moving along. In **(5)** ________________ , each segment has a pair of **(6)** ________________ . The most familiar annelids are probably **(7)** ________________ .

Determine if the statement is true. If it is not, rewrite the italicized part to make it true.

8. Earthworms have a *mouth with tiny teeth* in which food particles are ground up before entering the digestive tract. ________________________________

9. Blood is pumped throughout an earthworm's closed circulatory system by *an elongated, four-chambered heart.* ________________________________

10. Some body segments in annelids are *specialized for reproduction.* ________________________________

Below are the field notes of a biologist studying several newly collected annelid worms. Write the type of annelid—earthworm, bristleworm, or leech—being described.

11. Collected in rain forest of Papua, New Guinea; very active; flattened, with 32 body segments; has suckers on the ends of its body; no setae

Type of annelid: ________________________

12. Found crawling over corals on a reef; contains only eggs; no male reproductive organs; well-developed parapodia

Type of annelid: ________________________

13. Uncovered in top layer of moist soil; body has minute setae on ventral surface of each segment; hermaphroditic

Type of annelid: ________________________

Name Date Class

Chapter 28 Arthropods

Reinforcement and Study Guide

Section 28.1 Characteristics of Arthropods

In your textbook, read about what an arthropod is and exoskeletons.

Answer the following questions.

1. What is the most distinguishing arthropod characteristic?

____

2. Explain the advantage of having appendages with joints.

____

____

____

3. List three functions of an arthropod exoskeleton.

____

____

____

In your textbook, read about molting, segmentation, and gas exchange.

Complete each statement.

4. Prior to molting, a new exoskeleton forms ________________ the old one.

5. Many arthropods have three distinct body sections: a(n) ________________ , a(n) ________________ , and a(n) ________________ .

6. In arthropods that have a ________________ , the head and thorax are fused.

Complete the table by checking the correct column to indicate the respiratory structure you would expect to find in each example.

Example	Type of Respiratory Structure		
	Book Lungs	Gills	Tracheal Tubes
7. freshwater crayfish			
8. tarantula			
9. hissing cockroach			
10. swallowtail butterfly			

Chapter 28 **Arthropods,** *continued*

Reinforcement and Study Guide

Section 28.1 Characteristics of Arthropods, continued

In your textbook, read about arthropods' senses, body systems, and reproduction.

Identify the following as characteristics of either simple or compound eyes.

__________ **11.** have multiple lenses

__________ **12.** well-adapted for detecting slight movements

__________ **13.** have a single, focusing-type lens

__________ **14.** produce an image made up of thousands of parts

Determine if the statement is true. If it is not, rewrite the italicized part to make it true.

15. Animals produce pheromones, or *low frequency sounds,* that affect the behavior of others.

____

16. In many arthropods, large, fused ganglia act as nervous system control centers *for the entire body.*

____

17. Arthropods have an open circulatory system, in which blood leaves vessels and *comes in direct contact with body tissues.* __________

18. *Respiration* occurs in arthropods via the Malpighian tubules. __________

19. During parthenogenesis, *fertilized eggs* develop into offspring. __________

In your textbook, read about arthropod origins.

Answer the following questions.

20. What are the major reasons for the widespread success of arthropods?

____

____

21. From what animal group did arthropods probably evolve?

____

22. List three adaptations that have evolved in arthropods.

____

____

____

Chapter 28 **Arthropods,** *continued*

Section 28.2 Diversity of Arthropods

In your textbook, read about arachnids.

Circle the letter of the response that best completes the statement.

1. An animal that is *not* a member of the class Arachnida is
a. a spider. **b.** a deer tick. **c.** a walking stick. **d.** a dust mite.

2. In spiders, chelicerae are highly modified appendages that are adapted for
a. holding food and injecting poison. **b.** spinning silk and weaving webs.
c. chewing food. **d.** mating and reproduction.

3. The appendages of a spider that function as sense organs are
a. its chelicerae. **b.** its pedipalps. **c.** its legs. **d.** its spinnerets.

4. After catching their prey and injecting it with poison, spiders
a. eat the prey whole.
b. lay their eggs in the prey.
c. chew the prey into small pieces.
d. suck up the prey's contents, which have been liquified with enzymes.

5. In ticks and mites, the head, thorax, and abdomen
a. are absent. **b.** are well-defined.
c. are fused into one section. **d.** are all the same size.

6. The fact that horseshoe crabs have remained relatively unchanged for 500 million years indicates that
a. natural selection has not taken place. **b.** they must reproduce by parthenogenesis.
c. they have very little genetic diversity. **d.** their environment has changed very little.

In your textbook, read about crustaceans, centipedes, and millipedes.

Determine if each statement is true or false.

______________ **7.** Having compound eyes on movable stalks is an advantage for aquatic crustaceans whose potential predators could attack from almost any direction.

______________ **8.** The legs of most crustaceans are unspecialized and used only for walking.

______________ **9.** You might be more likely to see pill bugs moving around out in the open on a rainy day than on a sunny one.

______________ **10.** Both centipedes and millipedes have book lungs for gas exchange.

Section 28.2 Diversity of Arthropods, continued

In your textbook, read about insects.

Using the choices below, label the diagram of a honeybee.

antennae **compound eye** **legs** **mandibles** **spiracles** **wings**

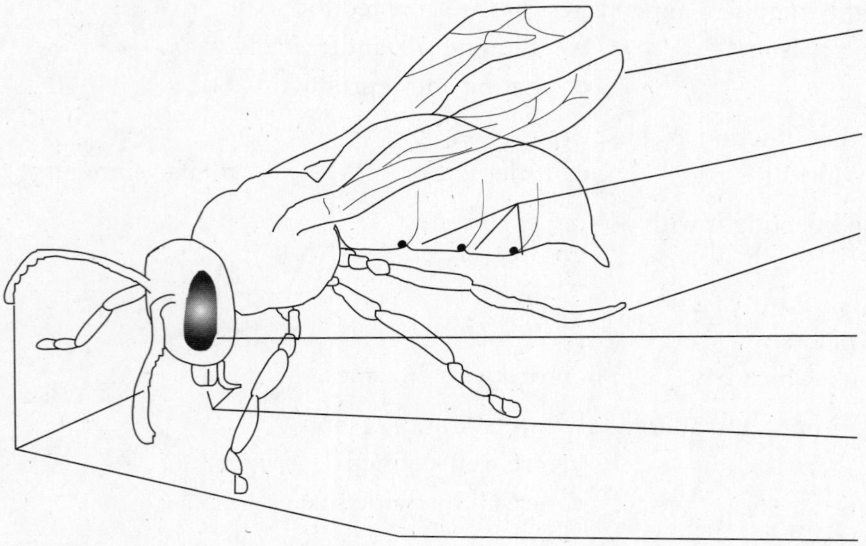

11. ______________

12. ______________

13. ______________

14. ______________

15. ______________

16. ______________

Complete the table by checking the correct column for each statement.

Description	Type of Metamorphosis	
	Complete	Incomplete
17. Insect begins life as a fertilized egg.		
18. Larva hatches from an egg.		
19. Nymph repeatedly molts and increases in size.		
20. Nymph hatches from an egg.		
21. Pupa undergoes changes while encased in cocoon.		
22. Adults and young usually eat the same food.		
23. Adults are the only sexually mature form.		

Name ____________________ Date __________ Class __________

Chapter 29

Echinoderms and Invertebrate Chordates

Reinforcement and Study Guide

Section 29.1 Echinoderms

In your textbook, read about echinoderms' internal skeleton, radial symmetry, and the water vascular system.

Answer the following questions.

1. Describe the "spiny skin" that is a characteristic of echinoderms.

____

____

____

2. In what way is being radially symmetrical an advantage for adult echinoderms?

____

____

____

____

For each item in Column A, write the letter of the matching item in Column B.

	Column A	Column B
________	**3.** Has a flattened, immovable endoskeleton made up of fused plates	**a.** brittle star
________	**4.** Has thin, flexible rays made up of small, overlapping, calcified plates	**b.** sea star
________	**5.** Has a flexible endoskeleton divided into rather long, tapering rays	**c.** sand dollar
________	**6.** Has tiny, calcified plates embedded in fleshy skin	**d.** sea lily
________	**7.** Has feathery, branching rays made up of tiny, calcified plates	**e.** sea cucumber

Complete the following sentences.

8. Tube feet are part of an echinoderm's ________ ________ ________, which is involved not only in locomotion, but also in ________ ________, ________, and food collecting.

9. In a sea star, water enters and exits the water vascular system through a structure called the ________, a sievelike, disc-shaped opening on the ________ side of the body.

Echinoderms and Invertebrate Chordates, *continued*

Section 29.1 Echinoderms, continued

In your textbook, read about sea star structure, echinoderm larvae, nutrition, nervous systems, and origins.

Label this drawing of a sea star and of a cross section of one of its rays. Use these choices:

ampulla **eyespot** **madreporite** **pedicellariae** **tube foot**

10. ______________________

11. ______________________

12. ______________________

13. ______________________

14. ______________________

Identify each of the following as describing either larva or an adult echinoderm.

______________ **15.** free-swimming

______________ **16.** bilaterally symmetrical

______________ **17.** radially symmetrical

______________ **18.** moves with tube feet

Determine if each of the following statements is <u>true</u> or <u>false</u>.

______________ **19.** If a sea urchin population underwent a population explosion, you might expect to see a rapid decline in the amount of algal life in the area.

______________ **20.** Sea stars and brittle stars both eat suspended organic particles.

______________ **21.** Most echinoderms have highly developed sense organs.

______________ **22.** The fact that echinoderms have bilaterally symmetrical larvae and deuterostome development is strong evidence that they are most closely related to chordates.

In your textbook, read about the diversity of echinoderms.

Answer the following questions.

23. List the five classes of living echinoderms and the types of animals in each class.

____

____

____

24. How is the ability to regenerate lost body parts adaptive for most echinoderms?

____

____

____

Complete the table by checking the column(s) that best fit(s) each description.

Description	Asteroidea	Ophiuroidea	Echinoidea	Holothuroidea	Crinoidea
25. Have multiple rays					
26. May rupture and release internal organs when threatened					
27. Some members of the class are sessile					
28. Burrow into rock or sand					
29. Use mucus-coated tentacles for feeding					
30. Some members of the class can actively swim from place to place					
31. Use rays, not tube feet, for locomotion					
32. The most inflexible type of echinoderm					
33. Use long, feathery arms to trap food particles drifting past					
34. Eat bivalves and other small animals					

Chapter 29

Echinoderms and Invertebrate Chordates, *continued*

Reinforcement and Study Guide

Section 29.2 Invertebrate Chordates

In your textbook, read about invertebrate chordates.

Complete the following sentences.

1. At some time in their life, all chordates possess a _______________, a dorsal hollow _______________ _______________, _______________ _______________, and muscle blocks.

2. During your early development, your notochord became your _______________, and your gill slits disappeared.

3. The _______________ _______________ is derived from the _______________ portion of the dorsal nerve cord, whereas the _______________ is derived from the anterior portion.

4. At some time during their lives, all chordates have a muscular _______________.

In your textbook, read about tunicates and lancelets.

Trace the path of water through a tunicate, starting with water entering the animal's body, by numbering the following statements from 1 to 5.

_______________ **5.** Water leaves the pharynx region.

_______________ **6.** Water passes through the gill slits, which filter food out of the water.

_______________ **7.** Water is drawn into the body through the incurrent siphon.

_______________ **8.** Water passes out of the body via the excurrent siphon.

_______________ **9.** Water enters the pharynx, where the gill slits are located.

Complete the table by checking the correct column(s) for each description.

Description	Tunicates	Lancelets
10. Only larval forms have a tail		
11. Are filter feeders		
12. Retain all chordate traits throughout life		
13. Blood flow is continually reversed in the adult body		
14. Capable of actively swimming as adults		

Name Date Class

BioDigest 8

Invertebrates

Reinforcement and Study Guide

In your textbook, read about invertebrates.

Study the definitions on the next page and write the terms in the appropriate spaces in the crossword puzzle below. All terms are important in the Biodigest.

1 2 3 4 5 6 7 8 9 10 11 12 13 14 15 16 17 18 19 20 21 22 23 24 25 26 27 28 29

ACROSS

1. In a animal with ____________ symmetry, the right and left sides are mirrors of each other.
7. An echinoderm has an inner skeleton; its ____________ covering is called the epidermis.
9. Bivalves, gastropods, and cephalopods are ____________ .
10. Some echinoderms have long ____________ that are used for locomotion.
12. The ____________ have radial symmetry and a water vascular system.
16. Because arthropods have ____________ , fossil arthropods are frequently found.
19. ____________ such as planaria have no body cavity.
21. In earthworms, internal ____________ are suspended from the mesoderm.
22. In some invertebrates, an exoskeleton offers ____________ and support for internal tissues.
25. A(n) ____________ belongs to the phylum Cnidaria.
26. Sponges have a(n) ____________ body shape.
27. In earthworms and other segmented worms, each ____________ has its own muscles.
28. Some segments in chordates have been modified into stacked layers called (2 words) ____________ .
29. In arthropods like grasshoppers, a set of jointed appendages called antennae are adapted to give the insect acute ____________ .

DOWN

1. The mouthparts of an arthropod may be adapted for such things as chewing, lapping, or __________ .
2. Echinoderm _________ have bilateral symmetry, which suggests a close relationship to the chordates.
3. ____________ go through metamorphosis during their life cycles.
4. A dorsal nerve ____________ is a hollow, fluid-filled canal lying above the notochord.
5. Setae, or small bristles, help earthworms with ____________ .
6. Mesoderm differentiates into ____________ , circulatory vessels, and reproductive organs.
8. Echinoderms have a supporting ____________ , which is inside of the body instead of outside.
11. The ____________ of a mollusk such as a clam is secreted by the mantle.
13. A(n) ____________ functions as a watery skeleton against which muscles can work.
14. Many ____________ are parasitic, such as *Trichinella*.
15. Arthropods are characterized by having a wide variety of ____________ for feeding.
17. The ____________ of a cnidarian is found in a highly specialized stinging cell.
18. ____________ are made up of two cell layers and have only one body opening.
20. Special ____________ feet enable sea stars to move from place to place.
22. Water enters a sponge through ____________ .
23. A jellyfish, a(n) ____________ , and an anemone are types of cnidarians.
24. Bivalves acquire food by filtering water through their ____________ .
26. A radula is a tonguelike organ used by snails to scrape ____________ from surfaces.

Chapter 30 **Fishes and Amphibians**

Reinforcement and Study Guide

Section 30.1 Fishes

In your textbook, read about what is a fish.

Complete each statement.

1. All vertebrates are in the phylum ________________, and fish, amphibians, reptiles, birds, and mammals are in the subphylum ________________.

2. Vertebrates are bilaterally symmetrical, coelomate animals with endoskeletons, closed ____________ systems, complex ________________, and efficient ________________ systems.

3. If you compared the number of fish species to the number of all other vertebrate species combined, there would be more species of ________________.

Complete the table below to compare the three different kinds of fishes.

Class	Kind of Fish	Jaws?	Skeleton	Fertilization
4. ____________	Lamprey and Hagfish	**5.** _____	**6.** ____________	**7.** ____________
Chondrichthyes	**8.** ____________	**9.** _____	Cartilage	**10.** ____________
11. ____________	**12.** ____________	Yes	**13.** ____________	External and Internal

Answer the following questions.

14. How does a fish breathe through its gills?

____

____

____

____

15. What two adaptations of cartilaginous and bony fishes help them to locate food?

____

____

____

In your textbook, read about bony fishes.

Determine if the statement is true. If it is not, rewrite the italicized part to make it true.

16. While spawning, a female bony fish may lay *millions of eggs* to be fertilized externally by the male, but only a few will survive. ______

17. The development of *bone* was an important event in the evolution of vertebrates because it eventually allowed them to move onto land and support their body weight. ______

18. The backbone, comprised of separate *gills*, was a major evolutionary event for fishes because it provided support as well as flexibility, which helped to propel them through the water. ______

19. The *swim bladder*, an organ found in bony fishes, allows fish to control their depth in water.

______

In your textbook, read about origins of fishes.

Examine the phylogenetic tree below. Then answer the question that follows.

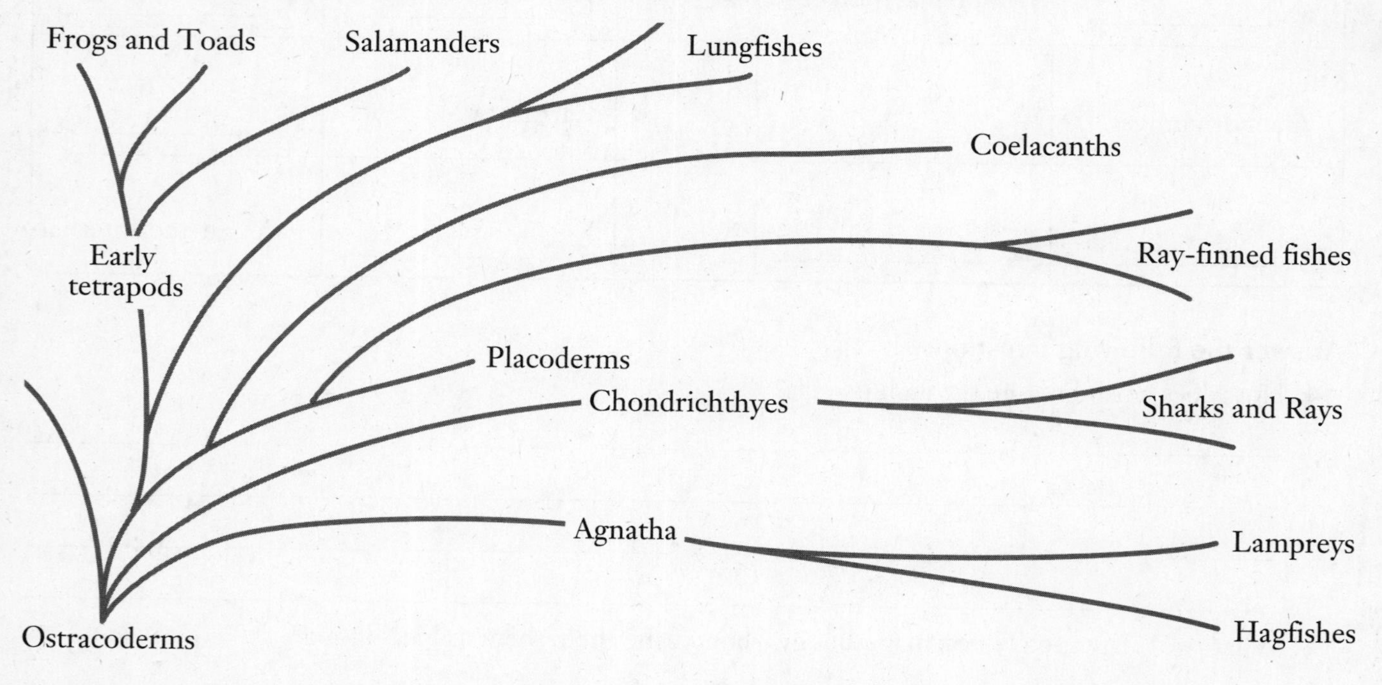

20. Which group of fishes do you think are most closely related to ancestral amphibians? Why?

______

______

______

Chapter 30 **Fishes and Amphibians,** *continued*

Reinforcement and Study Guide

Section 30.2 Amphibians

In your textbook, read about what is an amphibian.

Answer the following questions.

1. What three orders make up the class Amphibia?

____

____

____

2. Why do amphibian eggs need to be laid in water?

____

____

____

3. Where does an amphibian heart pump oxygen-rich blood, and where does it pump oxygen-poor blood?

____

____

____

____

For each item in Column A, write the letter of the matching item in Column B.

	Column A	Column B
________	**4.** Adult frogs and toads have legs, lungs, and a ________ heart.	**a.** two-chambered
________	**5.** Tadpoles have gills, fins, and a ________ heart.	**b.** three-chambered
________	**6.** Amphibians are ________ , animals whose body temperature changes with the temperature of their surroundings.	**c.** skin
________	**7.** Fertilized amphibian eggs hatch into ________ during the aquatic phase of their life.	**d.** tadpoles
________	**8.** Some salamanders have no lungs and breathe through their ________ .	**e.** ectotherms

Section 30.2 Amphibians, continued

In your textbook, read about the characteristics and diversity of amphibians.

Circle the letter of the response that best completes the statement.

9. Early amphibians needed large amounts of food and oxygen to
- **a.** walk on land.
- **b.** breathe on land.
- **c.** become dormant in cold weather.
- **d.** all of these.

10. In many amphibians, the most important organ for gas exchange is the
- **a.** blood.
- **b.** skin.
- **c.** lungs.
- **d.** circulatory system.

11. Many frogs and toads use ________ as a defense against predators.
- **a.** toxins
- **b.** electricity
- **c.** sharp claws
- **d.** all of these

12. Frogs and toads have sound-producing bands of tissues in their throat called
- **a.** tongues.
- **b.** vocal cords.
- **c.** vocal tissue.
- **d.** none of these.

13. Salamanders are unlike frogs and toads because they have
- **a.** long, slender bodies.
- **b.** tails.
- **c.** necks.
- **d.** all of these.

14. Caecilians are amphibians that have no
- **a.** eyes.
- **b.** skin.
- **c.** limbs.
- **d.** heart.

In your textbook, read about the origins of amphibians.

For each statement below, write true or false.

________________ **15.** Ancestral amphibians appeared on Earth about the same time as ancestral fishes.

________________ **16.** Amphibians probably evolved from tetrapods during the Paleozoic Era.

________________ **17.** Because the climate was hot and dry when amphibians first appeared on Earth, they had to stay near water.

________________ **18.** Like modern-day salamanders, early amphibians probably had legs set at right angles to the body.

________________ **19.** Because amphibians were a transitional group, they never were the dominant vertebrates on land.

Name Date Class

Chapter 31

Reptiles and Birds

Reinforcement and Study Guide

Section 31.1 Reptiles

In your textbook, read about what is a reptile and the amniotic egg.

Complete the following table about reptilian adaptations and their advantages by writing in the missing information in each case.

Adaptation	Advantage
1.	In crocodilians, oxygenated and deoxygenated blood kept separate; higher level of energy production
2. Thick, scaly skin	
3.	Water not necessary for fertilization
4. Legs positioned for walking and running on land	
5.	Water not necessary for reproduction; young not overly vulnerable to aquatic predators; prevents injury or dehydration of embryo

Label the diagram below, using these choices:

albumen allantois amnion chorion embryo shell yolk sac

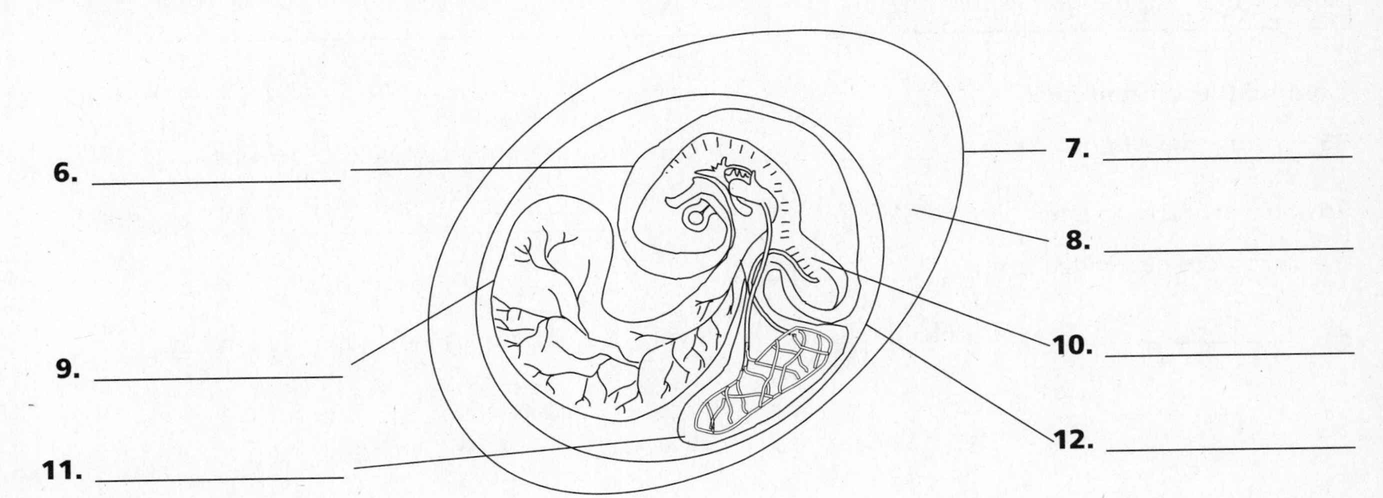

Chapter 31 **Reptiles and Birds,** *continued*

Reinforcement and Study Guide

Section 31.1 Reptiles, continued

In your textbook, read about the diversity of reptiles and the origins of reptiles.

Complete the chart by checking the correct column(s) for each characteristic.

Characteristic	Snakes	Lizards	Turtles	Crocodiles
13. Guard their nests against predators				
14. Possess shells				
15. Use tongue and Jacobsen's organ for smelling				
16. Kill prey by drowning it				
17. Lack limbs				
18. Have vertebrae and ribs fused to a carapace				
19. Some change color dramatically				
20. Lack teeth				
21. Some inject venom with fangs				
22. Some are aquatic				
23. Are primarily insect eaters				
24. Include marine species that migrate				

Complete each sentence.

25. During the Mesozoic era, ________________ were the most abundant land vertebrates.

26. Snakes and lizards are descended from early ________________ ________________, which in turn were descended from ________________.

27. ________________ are probably the modern, living descendants of some type of dinosaur.

Chapter 31 **Reptiles and Birds,** *continued*

Reinforcement and Study Guide

Section 31.2 Birds

In your textbook, read about what is a bird.

Answer the following questions.

1. From what type of animal are birds thought to have evolved?

2. List three physical features of birds that link them to reptilian ancestors.

3. Besides making flight possible, what other functions do feathers serve?

4. By what process are old feathers replaced?

In your textbook, read about how birds are adapted for flight.

Determine if the statement is true. If it is not, rewrite the italicized part to make it true.

5. A bird's sternum is *the point of attachment for its flight muscles.*

6. Being endothermic, birds have a body temperature that *fluctuates with environmental temperature.*

7. Because of its energy requirements, you might expect a bird to eat *less* than a reptile of comparable size.

8. Hollow bones, horny beaks, and a lack of teeth are all adaptations that make birds *more efficient predators.*

9. Birds grind up their food in a muscular *gizzard.*

10. The air inside a bird's lungs always has a fairly high *carbon dioxide* content, which makes for efficient gas exchange.

11. For a bird such as a goose or a duck, down feathers are the key to its superior *waterproofing.*

Section 31.2 Birds, continued

In your textbook, read about the diversity of birds.

For each item in Column A, write the letter of the matching item in Column B.

	Column A	Column B
______	**12.** long beak that is used for dipping into flowers to obtain nectar	**a.** owl
______	**13.** wings and feet modified for swimming; body surrounded with a thick layer of insulating fat	**b.** pelican
______	**14.** short, stout beak that is adapted to cracking seeds	**c.** hummingbird
______	**15.** feathered legs and feet that make it easier to walk in the snow	**d.** penguin
______	**16.** huge beak with a pouch that is used as a net for capturing fish	**e.** ptarmigan
______	**17.** large eyes, an acute sense of hearing, and sharp claws; adapted for nocturnal predation	**f.** goldfinch

In your textbook, read about the origins of birds.

Complete each statement.

18. The fossil record of birds is incomplete because bird skeletons are ______________ and ______________.

19. Unlike modern birds, ______________ had ______________, a long ______________, and ______________ ______________ ______________.

20. Scientists hypothesize that ______________ used its feathers for ______________, ______________, or ______________ ______________, rather than for flight.

Name Date Class

Chapter 32

Mammals

Reinforcement and Study Guide

Section 32.1 Mammal Characteristics

In your textbook, read about what is a mammal, and mammalian hair.

Answer the following questions.

1. Why are mammals able to live in almost every possible environment on Earth?

____

____

2. How do sweat glands help regulate body temperature?

____

____

Complete the table by checking the column that best fits each example.

Example	Serves As:		
	Camouflage	Defense	Warning
3. The striped fur of a tiger			
4. The sharp quills of a porcupine			
5. A skunk's black-and-white striped fur			
6. The white winter coat of an arctic hare			
7. The white hair patch on a pronghorn			

In your textbook, read about how mammals nurse their young, and about respiration and circulation.

Complete each statement.

8. Female mammals feed their young with milk produced by ____________ ____________.

9. In addition to milk and sweat, the glands of mammals produce ____________,

____________ ____________, and ____________.

10. The milk of mammals is rich in ____________, sugars, ____________, minerals, and

____________.

11. A mammal's muscular ____________ expands the ____________ ____________ bringing air into the lungs with each breath.

12. Like birds, mammals have ____________ hearts in which ____________ ____________ is kept entirely separate from ____________ ____________.

In your textbook, read about mammalian teeth, limbs, and learning.

Determine if the statement is true or false.

__________ **13.** The size and shape of a mammal's teeth can give valuable clues about its diet.

__________ **14.** Plant-eaters such as horses and cows have well-developed canine teeth for piercing food.

__________ **15.** The teeth of mammals are generally more uniform than the teeth of fishes and reptiles.

__________ **16.** By chewing their cud and then swallowing it, some mammals help bacteria break down the cellulose in their food.

__________ **17.** Mammalian limbs are adapted for a variety of methods of food gathering.

__________ **18.** Moles use their opposable thumbs to grasp objects.

__________ **19.** One reason mammals are successful is that they guard their young and teach them survival skills.

__________ **20.** Complex nervous systems and highly-developed brains make it possible for many kinds of mammals to learn.

Circle the letter of the response that best completes the statement.

21. Premolars and molars are used for
- **a.** shearing.
- **b.** crushing.
- **c.** grinding.
- **d.** all of these.

22. Cud chewing is an adaptation found in
- **a.** bears and other omnivores.
- **b.** tigers and other carnivores.
- **c.** many hoofed mammals.
- **d.** all of these.

23. The limbs of antelopes are characterized by
- **a.** greatly elongated finger bones.
- **b.** strong, slender bones.
- **c.** short bones and large claws.
- **d.** none of these.

24. Chimpanzees are intelligent enough to
- **a.** use tools.
- **b.** use sign language.
- **c.** work machines.
- **d.** all of these.

Name Date Class

Section 32.2 Diversity of Mammals

In your textbook, read about placental mammals, mammals with a pouch, and egg-laying mammals.

Answer the following questions.

1. What is a placental mammal?

____

____

2. What is the relationship between body size and gestation period in placental mammals?

____

3. Why are most marsupials found only in and around Australia?

____

____

____

4. What characteristic sets monotremes apart from all other mammals?

____

____

____

Complete the table by checking the correct column(s) for each characteristic.

Characteristic	Type of Mammal		
	Placental	**Marsupial**	**Monotreme**
5. Give birth to young			
6. Young nourished by a placenta during the entire development period			
7. Have a permanent pouch on abdomen			
8. Produce milk in mammary glands			
9. Lay eggs			
10. Include echidnas and duck-billed platypuses			
11. Have hair			
12. Comprise about 95 percent of all mammals			
13. Exhibit parental care			

In your textbook, read about the origins of mammals.

Use each of the terms below just once to complete the passage.

Cenozoic	**climate**	**dinosaurs**	**insect-eating**
mammals	**Pangaea**	**reptiles**	**therapsids**

Roughly 200 million years ago, the first **(14)** ________________ began roaming the earth. Most were small **(15)** ________________ descendants of **(16)** ________________, animals that had characteristics in common with both mammals and **(17)** ______________. The breakup of **(18)** ______________, the sudden disappearance of the **(19)** ________________, and a changing **(20)** ________________ provided early mammals with new food sources and habitats. During the **(21)** ________________ era, mammals diversified greatly into the many species inhabiting Earth today.

Examine the phylogenetic tree below. Then answer the questions that follow.

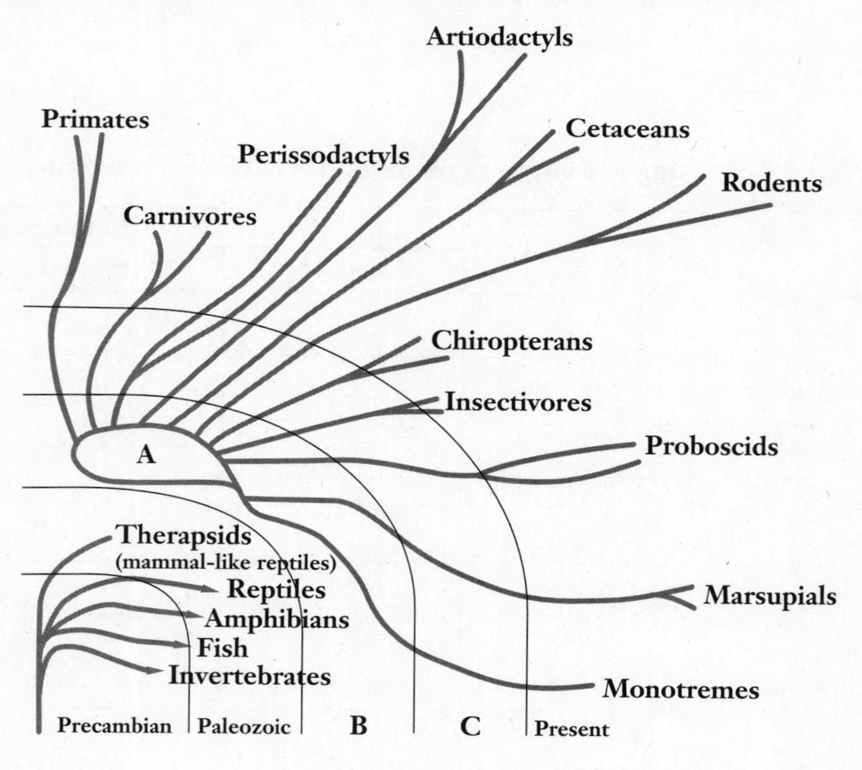

22. What group of animals is represented by the letter A above? ______________________________

23. What era is represented by the letter B? ______________________________

24. What era is represented by the letter C? ______________________________

Name ______________________ Date ______________ Class ______________

Chapter 33

Animal Behavior

Reinforcement and Study Guide

Section 33.1 Innate Behavior

In your textbook, read about what behavior is, inherited behavior, automatic responses to stimuli, and instinctive behavior.

Answer the following questions.

1. What is meant by animal behavior?

____

____

2. How is behavior adaptive?

____

____

3. Explain the relationship between innate behaviors and genetics.

____

____

____

4. What is an instinct?

____

Identify each of the following as being either a reflex or an instinct.

______________ **5.** You leap up after sitting down in shorts on a hot car seat.

______________ **6.** A sea turtle returns to the beach where she was hatched, in order to lay her eggs.

______________ **7.** A giant clam closes its shell when a shadow falls across it.

______________ **8.** A spider spins a complex, circular web.

In your textbook, read about courtship behavior and territoriality.

Determine if the following statements are true or false.

______________ **9.** Courtship behavior is something only male animals can instinctively perform.

______________ **10.** Courtship behavior is adaptive because it ensures that members of the same species can recognize each other and mate.

______________ **11.** A territory is a physical space that one animal defends against all other species of animals.

______________ **12.** Setting up territories reduces conflicts between members of the same species.

Chapter 33 **Animal Behavior,** *continued*

Reinforcement and Study Guide

Section 33.1 Innate Behavior, continued

In your textbook, read about aggressive behavior, submission, and behavior resulting from internal and external cues.

Below are excerpts from the field notebook of a behavioral biologist. Identify the behavioral phenomenon being described in each case.

13.

Field Notes

A large male baboon stares at another male and then suddenly "yawns" to reveal his long, sharp fangs.

Behavior Exhibited:

____

14.

Field Notes

When a herd of elephants arrives at a waterhole, the oldest female drinks first, followed by three females with calves, and finally a young male.

Behavior Exhibited:

____

15.

Field Notes

After fighting briefly with an older pack member, a young wolf stops fighting and rolls onto her back with her tail tucked between her legs and her eyes averted.

Behavior Exhibited:

____

16.

Field Notes

Large numbers of monarch butterflies fly south to roost in the winter.

Behavior Exhibited:

____

Complete the following sentences.

17. The type of dominance hierarchy formed by chickens is called a(n) _______________ _______________.

18. A cycle of behavior that occurs roughly every 24 hours is known as a(n) _______________ _______________.

19. Some animals use the positions of the _______________ _______________ _______________ to navigate. Others may use _______________ clues or Earth's _______________ _______________.

20. _______________ is similar to hibernation, in that metabolic activity _______________ in response to internal and external cues.

Chapter 33 **Animal Behavior,** *continued*

Reinforcement and Study Guide

Section 33.2 Learned Behavior

In your textbook, read about learned behavior.

Answer the following questions.

1. What is learned behavior?

2. What is a major advantage of being able to learn?

Complete the table by checking the correct column for each example.

Example	Type of Behavior	
	Learned	Innate
3. A dog catching a Frisbee		
4. A dog scent-marking a tree with urine		
5. A parrot saying "Polly want a cracker"		
6. A young lioness stalking prey with her mother		
7. A woodchuck going underground to hibernate		

In your textbook, read about habituation, imprinting, and learning by trial and error.

For each item in Column A, write the letter of the matching item in Column B.

	Column A	Column B
______	**8.** You stay with relatives who have a clock that chimes every hour. The first two nights, the chimes keep you awake, but after that you no longer notice them.	**a.** imprinting
______	**9.** A boy receives a day-old duckling as gift. It soon follows the boy wherever he goes.	**b.** trial-and-error learning
______	**10.** A young woman takes up archery. At first, her arrows don't hit the target, but after a week of practice, she is hitting the bull's eye fifty percent of the time.	**c.** habituation

Chapter 33 **Animal Behavior,** *continued*

Section 33.2 Learned Behavior, continued

In your textbook, read about conditioning and insight.

Determine if the statement is true. If it is not, rewrite the italicized part to make it true.

11. Learning by *repeating something over and over* is known as conditioning. ____________________

12. In Pavlov's conditioning experiments, the *innate reflex of salivating* was the stimulus that the dogs learned to associate with food. ____________________

13. Once conditioned, Pavlov's dogs would salivate at the sound of the bell *even when no food was present.*

____

14. A child figuring out how to use a chair to reach a cookie jar is an example of *trial and error.*

____

15. Insight is learning in which an animal uses *previous experience* to respond to a new situation.

____

In your textbook, read about the role of communication.

Use each of the terms below just once to complete the passage.

behavior	**communication**	**information**	**innate**	**language**
meanings	**odors**	**pheromones**	**sounds**	**symbols**

Through various forms of **(16)** ____________________, animals exchange **(17)** ____________________ that affects their **(18)** ____________________. Animals can communicate with visual signals, by touching each other, and by producing **(19)** ____________________, some of which can be heard over great distances. Communicating with **(20)** ____________________ is another strategy; **(21)** ____________________ are species-specific odor chemicals that can have a powerful effect on behavior. Some types of communication involve both **(22)** ____________________ and learned behavior. Human **(23)** ____________________ has evolved as a way of communicating with written and spoken **(24)** ____________________ that have specific **(25)** ____________________.

Name Date Class

BioDigest 9

Reinforcement and Study Guide

Vertebrates

In your textbook, read about fishes, amphibians, reptiles, birds, and mammals.

FISHES

Complete the chart by checking the correct column(s) for each characteristic of fishes.

Adaptation	Jawless	Cartilaginous	Bony
1. Jaws			
2. Gills			
3. Lateral line system			
4. Paired fins			
5. Skeleton made of cartilage			
6. Swim bladder			

AMPHIBIANS

Complete the following sentences.

7. Amphibians are ________________, which means that their body temperature depends upon the temperature of their surroundings. These vertebrates also carry out gas exchange through their ________________. Amphibians live on ________________ but reproduce in ________________. Almost all amphibians go through ________________, a radical change between the form of the young and the form of the adult.

REPTILES

Complete the table by describing the advantages that reptiles have because of certain adaptations. List one advantage for each adaptation.

Adaptation	Advantage
8. Scaly skin	
9. Amniotic egg	

BIRDS

Flight affects almost every system in birds. Explain the flight adaptations in each system listed here.

System	Adaptation
10. Bones	
11. Respiration	
12. Body covering	
13. Legs	
14. Wings	

MAMMALS

Various adaptations of mammals serve certain functions. In the space provided, write the letter of the adaptations that perform the function. Any letter may be used more than once.

Function

15. protection from low temperatures ______

16. protection from high temperatures ______

17. feeding young ______

18. stabbing or holding food ______

19. grinding or chewing food ______

20. providing large amounts of oxygen ______

Adaptation

a. hair
b. sweat glands
c. four-chambered heart
d. diaphragm
e. canine teeth
f. mammary glands
g. small ears
h. body fat
i. molars and premolars
j. hibernation
k. estivation

Name Date Class

Chapter 34

Protection, Support, and Locomotion

Reinforcement and Study Guide

Section 34.1 Skin: The Body's Protection

In your textbook, read about the structure and function of the skin.

Complete the table by checking the correct column for each description.

Description	Epidermis	Dermis
1. The outermost layer of skin		
2. Contains connective tissue, glands, and muscles		
3. The thicker, inner layer of skin		
4. Partly composed of dead, keratin-containing cells		
5. Contains pigmented cells that protect against the sun's rays		
6. Hair follicles grow out of this layer		
7. Site of continual mitotic cell divisions		
8. Richly supplied with blood vessels and nerves		

Answer the following questions.

9. Describe the change that takes place in your skin when you get a suntan.

10. How does skin help regulate body temperature?

11. List three other functions of skin.

Chapter 34

Protection, Support, and Locomotion, *continued*

Reinforcement and Study Guide

Section 34.2 Bones: The Body's Support

In your textbook, read about the structure of the skeletal system and joints.

Identify the following as being part of the axial or appendicular skeleton.

______________ **1.** the tarsals, metatarsals, and phalanges in your foot

______________ **2.** the seven vertebrae in your neck

______________ **3.** your rib cage

______________ **4.** the bones in your shoulder

______________ **5.** your lower jaw

______________ **6.** the humerus in your arm

For each answer below, write an appropriate question.

7. Answer: They are bands of connective tissue that attach muscles to bones.

Question: ____

8. Answer: It absorbs shocks and reduces friction between bones in a joint.

Question: ____

9. Answer: They connect bones to other bones.

Question: ____

10. Answer: One allows the bones to move back and forth; the other allows the bones to rotate.

Question: ____

In your textbook, read about the formation of bone and bone growth.

Complete each sentence.

11. In a human embryo's skeleton, ______________ is gradually replaced by ______________ except in a few places like the tip of the ______________.

12. Some cells in cartilage are stimulated to become ______________. They secrete a substance in which ______________ ______________ and other minerals are deposited.

13. Your bones increase in length near their ______________.

14. Even after you reach your full adult height, the bone-forming cells in your body will still be involved in ______________ and ______________.

In your textbook, read about compact and spongy bone and skeletal system functions.

Answer the following questions.

15. If you cut through to the center of a large leg bone, what bone components (in order, from the outside in) would you encounter?

____

____

16. How do blood vessels and nerves reach individual bone cells in compact bone?

____

17. What role does bone marrow play in the functioning of your circulatory system?

____

____

18. In what way is the skeleton a storehouse?

____

____

____

In your textbook, read about growth, mineral storage, and injury and disease in bone.

Determine if the statement is true or false.

__________ **19.** Once you have finished growing, your bones no longer change.

__________ **20.** Calcium is both deposited in and removed from bones.

__________ **21.** Calcium removed from bone is rapidly excreted in the urine as an unnecessary body waste.

__________ **22.** As a person ages, his or her bone density usually decreases.

__________ **23.** Because bones in an adult's skeleton are harder than children's bones, adults are less likely to break a bone in a fall.

__________ **24.** Osteoporosis is most common in older women because they rarely include milk in their diet.

Chapter 34

Protection, Support, and Locomotion, *continued*

Reinforcement and Study Guide

Section 34.3 Muscles for Locomotion

In your textbook, read about three types of muscles and skeletal muscle contraction.

Complete the table by checking the correct column for each description.

Description	Type of Muscle		
	Smooth	Skeletal	Cardiac
1. under voluntary control			
2. striated			
3. slow, prolonged contractions			
4. attached to bones			
5. found only in the heart			
6. not under voluntary control			
7. lines cavities and surrounds organs			

In your textbook, read about muscle strength and exercise.

Determine if the statement is true. If it is not, rewrite the italicized part to make it true.

8. Muscle strength depends on *the number of fibers in a muscle.*

____

9. When oxygen is limited, *aerobic respiration* becomes a muscle's primary source for ATP.

____

10. During anaerobic respiration, *oxygen* builds up in muscle cells.

____

11. A drop in the amount of lactic acid in the bloodstream indicates that muscular activity has *decreased.*

____

Chapter 35

The Digestive and Endocrine Systems

Reinforcement and Study Guide

Section 35.1 Following Digestion of a Meal

In your textbook, read about the functions of the digestive tract, the mouth, and the stomach.

Complete each statement.

1. The entire process of digestion involves first ____________________ food, then ____________________ it into simpler compounds, then ____________________ nutrients for use by body cells, and, finally, ____________________ wastes.

2. By chewing your food, you ____________________ its surface area.

3. Various enzymes play a role in ____________________ digestion, while the action of teeth, tongue, and muscles are involved in ____________________ digestion.

4. In your mouth, the enzyme ______________ is released from ______________ glands to begin the chemical breakdown of ____________________ .

5. Your ____________________ are adapted for cutting food, while your ____________________ are best suited for grinding food.

Determine if the statement is true. If it is not, rewrite the italicized part to make it true.

6. During swallowing, the epiglottis covers the *esophagus* to prevent choking.

____

7. Food is moved through the digestive tract by rhythmic waves of *voluntary muscle contractions* called peristalsis.

____

8. The churning actions of the stomach help mix the food with *pancreatic juices*.

____

9. Pepsin is a *protein-digesting enzyme* that only works in an acidic environment.

____

10. The stomach releases its contents into the small intestine *suddenly, all at once*.

____

Chapter 35

The Digestive and Endocrine Systems, *continued*

Reinforcement and Study Guide

Section 35.1 Following Digestion of a Meal, continued

In your textbook, read about the small intestine and the large intestine.

Answer the following questions.

11. What role do the enzymes secreted by the pancreas play in the digestive process?

12. Explain the relationship between the liver, the gallbladder, and bile.

13. Once in the small intestine, what happens to

a. digested food?

b. indigestible materials?

Complete the table by checking the correct column(s) for each function.

Function	Small Intestine	Large Intestine
14. Water is absorbed through walls.		
15. Digestion is essentially completed.		
16. Vitamin K is produced.		
17. Nutrients are absorbed by villi.		
18. Contents are moved by peristalsis.		
19. Indigestible material is collected.		
20. Bile and pancreatic juices are added.		

Chapter 35

The Digestive and Endocrine Systems, *continued*

Reinforcement and Study Guide

Section 35.2 Nutrition

In your textbook, read about carbohydrates, fats, and proteins.

Complete the table by checking the correct column(s) for each description.

Description	Carbohydrates	Fats	Proteins
1. the most energy-rich nutrients			
2. sugars, starches, and cellulose			
3. broken down into amino acids			
4. part of a nutritious, balanced diet			
5. normally used for building muscle, but can be used for energy			
6. broken down into glucose, fructose, and other simple sugars			
7. used to insulate the body from cold			

In your textbook, read about minerals and vitamins, water, and metabolism and calories.

Complete each statement.

8. ______________________________ are inorganic substances that help to build tissue or take part in chemical reactions in the body.

9. Unlike minerals, ______________________________ are organic nutrients that help to regulate body processes.

10. The two major vitamin groups are the ______________________________ and the ______________________________ vitamins.

11. The energy content of food is measured in ______________________________ , each of which is equal to ______________________________ calories.

12. Despite the claims of many fad diets, the only way to lose weight is to ______________________________ more calories than you ______________________________ .

Chapter 35

The Digestive and Endocrine Systems, *continued*

Reinforcement and Study Guide

Section 35.3 The Endocrine System

In your textbook, read about control of the body and negative feedback control.

Complete each statement.

1. Internal control of the body is handled by the ______________________________ system and the ______________________________ system.

2. Most endocrine glands are controlled by the action of the ______________________________ , or master gland.

3. A(n) ______________________________ is a chemical released in one part of the body that affects another part.

4. The amount of hormone released by an endocrine gland is determined by the body's ______________________________ for that hormone at a given time.

5. A ______________________________ system is one in which hormones are fed back to inhibit the original signal.

6. When your body is dehydrated, the pituitary releases ADH hormone, which reduces the amount of ______________________________ in your urine.

7. When you have just eaten and your blood glucose levels are high, your pancreas releases the hormone ______________________________ , which signals the liver to take in glucose, thereby lowering blood glucose levels.

In your textbook, read about hormone action, adrenal hormones and stress, and other hormones.

For each item in column A, write the letter of the matching item from Column B.

	Column A	Column B
________	**8.** Determines the body's food intake requirements	**a.** steroid hormones
________	**9.** Made from lipids and diffuse freely into cells through the plasma membrane	**b.** glucocorticoids and aldosterone
________	**10.** Bind to receptors embedded in the plasma membrane of the target cell.	**c.** calcitonin and parathyroid hormone
________	**11.** Produce a feeling called "adrenaline rush"	**d.** epinephrine and norepinephrine
________	**12.** Help the body prepare for stressful situations	**e.** amino acid hormones
________	**13.** Regulate calcium levels in blood	**f.** thyroxine

Name Date Class

Chapter 36

The Nervous System

Reinforcement and Study Guide

Section 36.1 The Nervous System

In your textbook, read about neurons–basic units of the nervous system.

Complete the table by filling in the missing information in each case.

Structure	Function
1.	carry impulses toward the brain and spinal cord
2. dendrites	
3. motor neurons	
4.	transmit impulses within the brain and spinal cord
5.	carry impulses away from neuron cell bodies

Order the steps in impulse transmission from 1 to 7.

_______ **6.** A wave of depolarization moves down the neuron.

_______ **7.** The Na^{+}/K^{+} pump takes over again, pumping sodium ions out across the membrane, and pumping potassium ions in.

_______ **8.** Sodium channels in the neural membrane open.

_______ **9.** A neuron receives a stimulus.

_______ **10.** As the wave of depolarization passes, sodium channels close and potassium channels open.

_______ **11.** The neuron returns to a resting state.

_______ **12.** Sodium ions flow into the neuron, causing the inside of the neuron to become positively charged.

Chapter 36 **The Nervous System,** *continued*

Reinforcement and Study Guide

Section 36.1 The Nervous System, continued

In your textbook, read about the central nervous system and the peripheral nervous system.

Label the diagram of the brain to show the cerebrum, cerebellum, and brain stem.

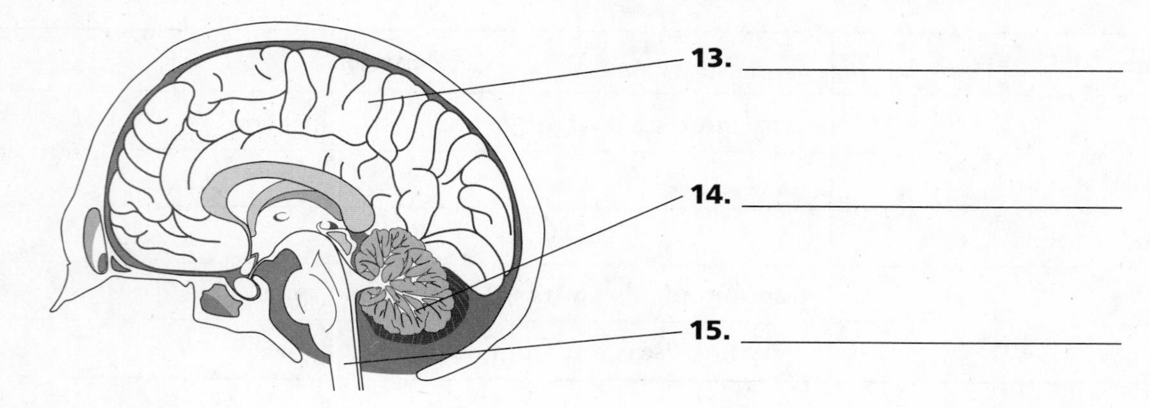

Write the name of the part labeled above that matches each description in the table.

Description	Part
16. Includes the medulla and pons	
17. Controls conscious activities and movement	
18. Important for keeping your balance	
19. If damaged, heart rate might be affected	
20. If damaged, memory might be affected	
21. Ensures that movements are coordinated	

Complete the table by checking the correct column for each description.

	Autonomic Nervous System Division	
Description	**Sympathetic**	**Parasympathetic**
22. Controls internal activities when the body is at rest		
23. Increases breathing rate		
24. Tenses muscles		
25. Slows heart rate down		
26. Activates fight or flight response		

Chapter 36 **The Nervous System,** *continued*

Section 36.2 The Senses

In your textbook, read about sensing chemicals and sensing light.

Determine if each statement is true or false.

__________ **1.** Impulses coming from sensory receptors in your nose and mouth are interpreted as odors and tastes by the cerebrum.

__________ **2.** All of your tongue's tastebuds respond equally well to all taste sensations.

__________ **3.** The lens in the eye controls the amount of light that strikes the retina.

__________ **4.** On a bright sunny day, the cones in your eyes play a greater role in your sense of sight than the rods.

__________ **5.** Only the left hemisphere of the brain is involved in the sense of sight.

__________ **6.** When you are looking at an object, each of your eyes sees the object from the same perspective.

__________ **7.** Much of what you taste depends on your sense of smell.

In your textbook, read about sensing mechanical stimulation.

Circle the letter of the response that best completes each statement.

8. Sound waves are converted into nerve impulses inside the
a. ear canal. **b.** cochlea. **c.** malleus. **d.** optic nerve.

9. If the semicircular canals in one of your ears were damaged, you might
a. lose your ability to hear low-frequency sounds.
b. lose your ability to coordinate your neck muscles.
c. lose your sense of balance.
d. lose your sense of rhythm.

10. The malleus, incus, and stapes are found in the
a. outer ear. **b.** eardrum. **c.** middle ear. **d.** inner ear.

11. Your senses of hearing and touch *both* depend on nerve impulses being generated by
a. electrical stimulation. **b.** sound waves.
c. a change in temperature. **d.** mechanical stimulation.

12. In the skin of your fingertips, you might expect to find receptors for
a. touch. **b.** pressure. **c.** pain. **d.** all of these

Chapter 36 **The Nervous System,** *continued*

Reinforcement and Study Guide

Section 36.3 The Effects of Drugs

In your textbook, read about how drugs act on the body, their medicinal uses, and abuse of drugs.

Answer the following questions.

1. Distinguish between a drug and a medicine.

2. What is a narcotic?

3. Compare the effect of a stimulant on the CNS with the effect of a depressant.

4. What is an addiction?

5. How does a person's body develop a tolerance for a drug?

In your textbook, read about the classes of commonly abused drugs.

Complete the table by checking the correct column for each example.

Example	Stimulant	Depressant
6. Drugs that cause an increase in heart rate		
7. Alcohol		
8. Nicotine		
9. Drugs that increase neurotransmitter levels		
10. Barbiturates		
11. Drugs that cause vasoconstriction		
12. Opiates		
13. Hallucinogens		

Name Date Class

Chapter 37

Respiration, Circulation, and Excretion

Reinforcement and Study Guide

Section 37.1 The Respiratory System

In your textbook, read about air passageways and lungs.

Circle the letter of the choice that best completes the statement or answers the question.

1. During the process of respiration,
a. oxygen is delivered to body cells.
b. carbon dioxide is expelled from the body.
c. oxygen is used in cells to produce ATP.
d. all of these.

2. When you swallow, your epiglottis momentarily covers the top of the trachea so that
a. you can swallow more easily.
b. you can breathe more easily.
c. you don't get food in your air passages.
d. you can cough up foreign matter.

3. The cilia that line your trachea and bronchi
a. produce dirt-trapping mucus.
b. help in the exchange of oxygen and CO_2.
c. move mucus and dirt upward.
d. only beat when you inhale.

4. The first branches off the trachea are called
a. bronchioles. **b.** bronchi. **c.** arterioles. **d.** alveoli.

5. Inside the alveoli, carbon dioxide and oxygen
a. are exchanged between air and blood.
b. are transported along microscopic tubules.
c. are produced inside cells.
d. are exchanged for other gases.

6. Which is the correct sequence for the path of oxygen through the respiratory system?
a. nasal passages, bronchi, trachea, bronchioles, cells, blood, alveoli
b. cells, blood, alveoli, bronchioles, bronchi, trachea, nasal passages
c. nasal passages, blood, alveoli, bronchi, cells, trachea, bronchioles
d. nasal passages, trachea, bronchi, bronchioles, alveoli, blood, cells

In your textbook, read about the mechanics of breathing and the control of respiration.

For each statement below, write true or false.

____________ **7.** Homeostasis in respiration is controlled by the cerebrum.

____________ **8.** As you exhale, the bronchioles in the lungs release most of their air.

____________ **9.** When you inhale, the muscles between your ribs contract.

____________ **10.** Relaxation of the diaphragm causes a slight vacuum in the lungs.

____________ **11.** Air rushes into the lungs because the air pressure outside the body is greater than the air pressure inside the lungs.

____________ **12.** Relaxation of the diaphragm causes it to flatten.

Chapter 37

Respiration, Circulation, and Excretion, *continued*

Reinforcement and Study Guide

Section 37.2 The Circulatory System

In your textbook, read about your blood, ABO blood types, and blood vessels.

Answer the following questions.

1. What cells and substances would you expect to find suspended or dissolved in plasma?

____

____

2. How is carbon dioxide transported in blood?

____

____

Complete the table below by checking the correct column for each description.

Description	Red Blood Cells	White Blood Cells	Platelets
3. Contain hemoglobin			
4. Fight infection			
5. Lack a nucleus			
6. Help clot blood			
7. Transport oxygen			
8. Comparatively large and nucleated			

For each statement below, write <u>true</u> or <u>false</u>.

_______________ **9.** Your blood type can be changed with a blood transfusion.

_______________ **10.** Different blood types result from different antibodies being present on the membranes of red blood cells.

_______________ **11.** If you have type B blood, then you have anti-A antibodies in your plasma.

_______________ **12.** Risks involving incompatible Rh factors are greatest for a woman's first child.

In your textbook, read about your heart, blood's path through the heart, and inside your heart.

Label the parts of the human heart in the diagram below. Use these choices:

aorta	left atrium	left ventricle	pulmonary arteries
pulmonary veins	right atrium	right ventricle	

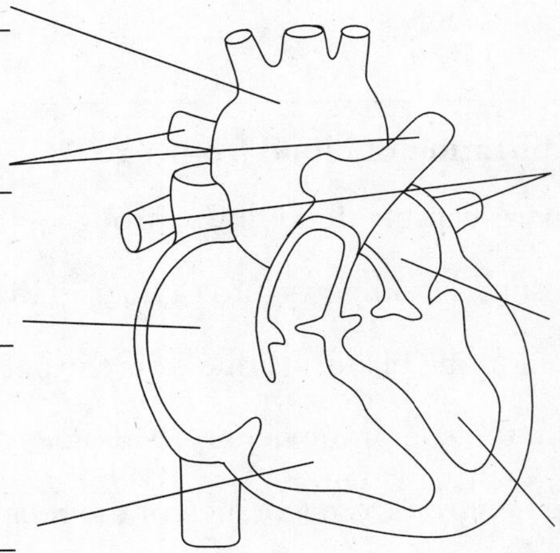

13. ____________

14. ____________

15. ____________

16. ____________

17. ____________

18. ____________

19. ____________

20. Where does blood go from the pulmonary veins? From the right ventricle? From the left ventricle?

____

21. What prevents blood from mixing between atria and ventricles?

____

In your textbook, read about heartbeat regulation, control of the heart, and blood pressure.

Determine if the statement is true. If it is not, rewrite the italicized part to make it true.

22. The surge of blood through an artery is called the *cardiac output*. ____________

23. The pacemaker initiates heartbeats *by generating electrical impulses*. ____________

24. An electrocardiogram is a record of *the strength of each heartbeat*. ____________

____

25. The *atrioventricular node*, along with sensory cells in arteries near the heart, regulates the pacemaker.

____

26. *Diastolic pressure* occurs when the heart's ventricles contract. ____________

Chapter 37

Respiration, Circulation, and Excretion, *continued*

Reinforcement and Study Guide

Section 37.3 The Urinary System

In your textbook, read about kidneys, nephrons, and the formation of urine.

Answer the following questions.

1. What is the major function of kidneys?

____

2. What role does the bladder play in the urinary system?

____

3. What are nephrons?

____

Order the following steps in the filtration of blood from 1 to 7.

________ **4.** From the Bowman's capsule, fluid flows through a U-shaped tubule.

________ **5.** Under high pressure, blood flows into capillaries that make up the glomerulus.

________ **6.** After being stored in the bladder, urine exits the body via the urethra.

________ **7.** Fluid moves from the end of the nephron's tubule to the ureter.

________ **8.** Blood enters the nephron from a branch of the renal artery.

________ **9.** Water, glucose, amino acids, and ions are reabsorbed into the blood.

________ **10.** Water, glucose, amino acids, wastes, and other substances move from glomerular capillaries into a Bowman's capsule.

In your textbook, read about the urinary system and homeostasis.

Complete each statement.

11. ________________ and ________________ are two toxic nitrogenous wastes that your kidneys constantly remove from your bloodstream.

12. The kidneys also help regulate the blood's ________________ ________________, and ________________.

13. Individuals with diabetes have excess levels of ________________ in their blood.

Chapter 38

Reproduction and Development

Reinforcement and Study Guide

Section 38.1 Human Reproductive Systems

In your textbook, read about human male anatomy and hormonal control.

Answer the following questions.

1. What are the primary functions of the male reproductive system?

____

____

2. How does the location of the scrotum affect sperm?

____

____

3. How many sperm can the average mature male produce in one day?

____

Order the steps in the formation and transportation of sperm from 1 to 6.

__________ **4.** Mature sperm enter the vas deferens.

__________ **5.** Newly formed haploid sperm cells pass through a series of coiled ducts to the epididymis.

__________ **6.** Sperm leave the body via the urethra.

__________ **7.** Sperm mature in the epididymis.

__________ **8.** Cells lining tubules in the testes undergo meiosis.

__________ **9.** Sperm travel along the ejaculatory ducts and into the urethra.

Complete each sentence.

10. When a young man's "changes," he is probably entering ____________________, a time when he will develop other secondary ________________ ________________.

11. A hormone released by the ________________ stimulates the ________________ gland to release ________________ - ________________ and ________________ hormones.

12. FSH regulates ________________ production, while LH controls the production of the steroid hormone ________________ by the testes.

In your textbook, read about human female anatomy and puberty in females.

Determine if each statement is true. If it is not, rewrite the italicized part to make it true.

13. When an egg cell is released from an ovary, it moves down the oviduct by *gravity*.

____

14. As is the case in human males, a woman's *hypothalamus* produces FSH and LH.

____

15. *FSH* stimulates follicle development and the release of estrogen from the ovary.

____

16. In females, *luteinizing hormone (LH)* is responsible for the development of the secondary sex characteristics.

____

17. Long before a woman is born, cells in her ovaries that are destined to become future eggs undergo several *mitotic divisions*.

____

In your textbook, read about the menstrual cycle.

Complete the table by checking the correct column for each event.

Event	Phase of Menstrual Cycle		
	Flow	**Follicular**	**Luteal**
18. LH stimulates the corpus luteum to develop from a ruptured follicle.			
19. Estrogen levels are at their peak.			
20. A cell inside a follicle resumes meiotic divisions.			
21. Progesterone levels are at their peak.			
22. The uterine lining is shed.			
23. LH levels rise abruptly.			
24. Ovulation occurs.			
25. The uterine lining becomes engorged with blood, fat, and tissue fluid.			
26. FSH begins to rise.			

Section 38.2 Development Before Birth

In your textbook, read about fertilization and implantation, and embryonic membranes.

Use each of the terms below just once to complete the passage.

amnion	blastocyst	chorion	chorionic villi	embryo
implants	oviduct	placenta	umbilical cord	zygote

Usually in the upper part of a(n) **(1)** ______________________, an egg and one sperm unite to form a(n) **(2)** ______________________. This single cell divides repeatedly to form a(n) **(3)** ______________________, which **(4)** ______________________ in the uterine wall. Part of the blastocyst becomes the **(5)** ______________________, which is surrounded by a fluid-filled, membranous sac called the **(6)** ______________________. The embryo is connected to the wall of the uterus by its **(7)** ______________________. The amniotic sac is enclosed by the **(8)** ______________________, which later forms the **(9)** ______________________. Nutrients and oxygen from the mother and wastes from the embryo are exchanged in the **(10)** ______________________.

In your textbook, read about fetal development and genetic counseling.

Complete the table by checking the correct column for each event or example.

Event/Example	Trimester		
	First	**Second**	**Third**
11. Fetus can survive outside the uterus with medical assistance.			
12. Fetus weighs more than 3000 grams.			
13. Embryo is most vulnerable to outside influences.			
14. Embryo becomes a fetus.			
15. Fetus can use its muscles to move spontaneously.			
16. Fetus becomes oriented head-down.			
17. Sex of fetus can be determined.			

Chapter 38 **Reproduction and Development,** *continued*

Reinforcement and Study Guide

Section 38.3 Birth, Growth, and Aging

In your textbook, read about birth.

Answer the following questions.

1. What are the three stages of birth? _______________

2. Describe the action of oxytocin. _______________

____

3. After the placenta is expelled from a woman's body, what effect do continued uterine contractions have? _______________

____

In your textbook, read about growth and aging.

Complete each sentence.

4. Your growth rate, as well as the type of growth you undergo, varies with both your _______________ and your _______________.

5. _______________ _______________ _______________, abbreviated _______________, regulates growth.

6. hGH exerts its effects primarily on _______________ and on _______________.

7. LGH works by increasing _______________ and _______________.

Complete the table by checking the correct column for each description.

Example	Childhood	Adolescence	Adulthood
8. Your growth rate continues at a steady rate.			
9. Lines develop on your face, especially around your eyes and mouth.			
10. You reach maximum physical stature.			
11. You begin to reason.			
12. You may have a sudden growth spurt.			

Name Date Class

Chapter 39

Immunity From Disease

Reinforcement and Study Guide

Section 39.1 The Nature of Disease

In your textbook, read about what an infectious disease is, determining what causes a disease, and the spread of infectious diseases.

Answer the following questions.

1. Why is a disease like osteoarthritis not considered an infectious disease?

____

2. What is meant by Koch's postulates?

____

____

3. In terms of disease, what is a reservoir?

____

Complete the table by writing in the method of transmission for each example.

Example	Method of Transmission
4. While exploring a cave, a person breathes in fungal spores that cause a lung infection.	
5. A person contracts Rocky Mountain spotted fever after being bitten by a tick.	
6. After having unprotected sex, a person contracts syphilis.	

In your textbook, read about what causes the symptoms of a disease, patterns of disease, and treating diseases.

For each statement below, write true or false.

______________ **7.** The toxin produced by a particular microorganism might be far more destructive than the direct damage the microbe does to its host cells.

______________ **8.** Endemic diseases often disappear in a population, only to resurface unexpectedly many years later.

______________ **9.** If you catch the flu during an influenza epidemic, your best hope of recovery is to take antibiotics.

______________ **10.** It is important for researchers to try to discover new antibiotics because many types of bacteria are becoming resistant to the ones now being used.

Chapter 39 **Immunity From Disease,** *continued*

Reinforcement and Study Guide

Section 39.2 Defense Against Infectious Diseases

In your textbook, read about the innate immune system.

Determine if the statement is true. If it is not, rewrite the italicized part to make it true.

1. Healthy skin is a good defense against the invasion of pathogens because it is *free of bacteria*.

____

____

2. In your trachea, *saliva* traps microbes and prevents them from entering your lungs.

____

3. Macrophages migrate *into the bloodstream* when the body is challenged by a pathogen.

____

4. Phagocytes at the site of an infection or inflammation destroy pathogens by surrounding and engulfing them. ______________________________

5. The third line of defense against infection is the consumption of pathogens by *neutrophils*.

____

6. Interferon is produced by cells infected by *pathogenic bacteria*.

____

In your textbook, read about acquired immunity.

Circle the letter of the choice that best completes the statement.

7. The human lymphatic system is important in
a. filtering pathogens from lymph.
b. keeping body fluids constant.
c. resistance to disease.
d. all of the above.

8. Tissue fluid is found
a. in lymph vessels.
b. in the bloodstream.
c. around body cells.
d. in lymph ducts.

9. The main function of lymph nodes is to
a. store red blood cells.
b. filter lymph.
c. filter excess fluid.
d. trigger an immune response.

10. A reservoir for lymphocytes that can be transformed into specific disease-fighting cells is the
a. thymus gland.
b. thyroid gland.
c. pituitary gland.
d. pancreas.

In your textbook, read about antibody immunity and cellular immunity.

Complete each sentence.

11. ______________________ is the building up of a ______________________ to a specific pathogen.

12. Two types of immunity that involve different kinds of cells and cellular actions are ______________________ immunity and ______________________ immunity.

13. The presence of foreign ______________________ in the body triggers the production of ______________________ by plasma cells.

14. A ____________________ is a lymphocyte that, when activated by a ______________________, becomes a plasma cell and produces ______________________ .

15. Cellular immunity involves several different types of ______________________ cells.

16. A ____________________ releases enzymes directly into the ____________________ .

Complete the table by checking the correct columns for each example.

Example	Type of Immunity	
	Cellular	Antibody
17. Involves the protection of antibodies		
18. Simulated by antigens in the body		
19. Clones of killer T cells produced		
20. Memory cells produced so the body can respond quickly to a second attack		
21. Key role played by antigen-antibody complex		
22. T cells destroyed by pathogens directly		

Chapter 39 **Immunity From Disease,** *continued*

Reinforcement and Study Guide

Section 39.2 Defense Against Infectious Diseases, continued

In your textbook, read about passive and active immunity to infectious diseases.

Answer the following questions.

23. Distinguish between active and passive immunity.

____

____

____

24. In what two ways can passive immunity develop?

____

____

25. What is a vaccine?

____

____

In your textbook, read about AIDS and the immune system.

For each statement below, write true or false.

______________ **26.** The virus that causes AIDS—Human Immunodeficiency Virus—is well-named because it attacks the immune system.

______________ **27.** HIV can be transmitted by air.

______________ **28.** A child born to a woman who is infected with HIV is at high risk for being infected, too.

______________ **29.** HIV destroys a person's resistance to disease by attacking and destroying memory T cells.

______________ **30.** In a blood sample from an HIV-positive person, you would expect to find most of the viruses existing free in the blood, rather than being hidden inside cells.

______________ **31.** If a person is infected with HIV, he or she will usually develop AIDS within about a year.

______________ **32.** The cause of death for a person with AIDS usually is some type of infection that the body's weakened immune system can no longer fight off.

______________ **33.** The majority of persons infected with HIV will develop AIDS.

Name ______________________ Date ______________ Class ______________

BioDigest 10

The Human Body

Reinforcement and Study Guide

In your textbook, read about skin, bones, and muscles.

Skin has four functions: **(1)** ______________________, **(2)** ______________________, **(3)** ______________________, and **(4)** ______________________. These functions help maintain homeostasis in the body.

Complete the table to describe the role of bones.

Support for	**5.**
Place for	**6.**
Protects	**7.**
Manufacture of	**8.**
Storehouse of	**9.**

This diagram shows the various steps involved in the respiratory process. In the space provided, describe the steps as indicated.

Step 1: Oxygen enters the lungs when you inhale.

LUNG

Step 7: Carbon dioxide leaves the lungs when you exhale.

Step 2: **(10)** ______________________

Bloodstream

Bloodstream

Step 6: **(13)** ______________________

Step 3: Oxygen passes from blood to the cells.

Cell

O2 + Glucose

ATP

CO2

Step 5: **(12)** ______________________

Step 4: **(11)** ______________________

BioDigest **10** **The Human Body,** *continued*

Reinforcement and Study Guide

In your textbook, read about reproductive, endocrine, and lymphatic systems.

14. The endocrine system is a communication system; its messages are hormones. They are produced by ______________ and travel in the ______________ to ______________ ______________. There they ______________.

15. The structures of the male reproductive system—the scrotum, ______________, epididymis, seminal vesicles, ______________, bulbourethral gland, urethra, and ______________—are involved in ______________ and maintaining sperm cells and ______________ into the female reproductive tract.

16. The structures of the female reproductive system—the ______________, oviduct, ______________, and vagina—produce and maintain ______________, receive and transport ______________, and support the development of the ______________.

Explain how these systems of the body interact with each other.

17. Skeletal system ⟶ Circulatory system ⟶ Muscular system

____

____

18. Digestive system ⟶ Circulatory system ⟶ Urinary system

____

____

19. Endocrine system ⟶ Reproductive system

____

____

20. Respiratory System ⟶ Circulatory system ⟶ Cell ⟶ Circulatory system ⟶ Urinary and Respiratory systems

____

____

____

21. Circulatory system ⟶ Lymphatic system

____

____